Composite reinforced concrete

Composite reinforced concrete

R. TAYLOR
BSc, DIC, MICE, MIStructE
Senior Lecturer in Civil Engineering
University of Manchester
Simon Engineering Laboratories

Thomas Telford Limited, London, 1979

Published by Thomas Telford Limited, Telford House, PO Box 101,
26–34 Old Street, London EC1P 1JH

ISBN: 0 7277 0077 4

© R. Taylor, 1979

Typeset and printed by Henry Ling Limited, Dorchester

FOREWORD

There is no doubt that none of the major structural materials is ideal. Despite the long-standing use of various combinations of steel and concrete, all methods of combining these materials have some drawbacks. While ordinary reinforced concrete combines the use of the tensile strength of steel with the compressive strength of concrete, it cannot take full advantage of high tensile steels because of the limited capacity of concrete to undergo tensile strain without excessive cracking. Although this problem is overcome in prestressed concrete, the economy resulting from the use of ultra-high strength steels is offset by the costly prestressing operation, at least for site construction. The other main traditional steel—concrete combination — composite construction — has undoubted advantages in construction but is extravagant in its use of steel.

It would be advantageous if the economic use of ultra high strength tensile steel could be combined with the convenience of composite construction. While composite reinforced concrete does not completely achieve this ideal, it goes some way towards it.

The Author of this monograph, Mr R. Taylor, has developed his proposals for composite reinforced concrete as a result of pursuing two other lines of research. The first arose from the aim of improving the efficiency of the use of steel in composite construction by incorporating deep haunches in the beams without sacrificing constructional advantages. In the second investigation, Mr Taylor sought a means of improving the exploitation of reinforced concrete beyond the limits currently recommended by Codes of Practice. As a result of these two investigations, it was realized that the incorporation of deep haunches in composite beams could be achieved alongside the use of high tensile reinforcement to give a possible improvement in overall economy.

It is hardly likely that innovatory methods of construction will be accepted unless they show distinct advantages. Composite reinforced concrete may well have such advantages in multi-storey structures when it is desired to use precast flooring units in conjunction with an in situ concrete frame in which the beams need to be of minimum constructional depth.

The present monograph is published with the intention of stimulating practical exploration of this proposed method of construction. Considerable theoretical economies are possible; the realization of these economies in practice could lead to major steps forward in the more economic exploitation of steel—concrete composite construction.

M. R. Horne
Manchester
March 1979

PREFACE

This research monograph on composite reinforced concrete is based on lectures given at a research seminar in April 1979 at the Simon Engineering Laboratories of the University of Manchester. At this stage composite reinforced concrete is only a research material and has not been used in practice. Indeed hitherto it has not been possible to use composite reinforced concrete as there has been insufficient information available to the designer. It was the purpose of the seminar, and is the purpose of this monograph, to make available the information that has been obtained at the University of Manchester from research over a period of several years, and also to indicate those areas in which more information is still needed.

Although aimed essentially at engineers concerned with research in concrete structures, a wide cross-section of engineers from industry also attended the seminar. Such varying backgrounds of the participants of the seminar determined the material presented and its manner of presentation. It could not be assumed, for example, that each had a good knowledge of reinforced concrete design and the requirements of CP 110, nor that each had a good knowledge of composite construction and the requirements of CP 117. For this reason somewhat fuller explanations are given than would normally be the case in a research report.

Chapter 1 introduces the idea of composite reinforced concrete and presents examples of design for the case of simply supported beams. Chapter 2 discusses the problem of continuity and gives examples of design of continuous beams. Chapter 3 covers the problems of vertical and horizontal shear and the design of the shear connections. Chapter 4 discusses some miscellaneous problems such as deflexion, cracking, fatigue and fire resistance.

CONTENTS

CHAPTER 1

Simply supported beams

The combination of steel and concrete for structural purposes normally falls into one of the traditional categories, such as reinforced concrete, prestressed concrete and composite construction. These are quite well defined categories but it is not unknown for the boundaries to be crossed and prestressed reinforced concrete (or partially prestressed concrete) and prestressed composite construction are examples of steel—concrete combinations where the traditional boundaries have been crossed. All these various modes of construction have their advantages and disadvantages which determine their use in practice.

It will be suggested that a new combination of steel and concrete combines many of the advantages of the traditional forms whilst eliminating some of their attendant disadvantages. Nevertheless it is emphasized that this new material, designated composite reinforced concrete, is not suggested as a replacement of the traditional forms, but merely as an additional and alternative form to be used in appropriate circumstances. It is considered that composite reinforced concrete will be found to be most suitable for the multi-storey framed structure.

Starting with a consideration of costs, it is shown that the traditional forms for concrete structures leave something to be desired. It will be shown for example that, although reinforced concrete is very efficient in its use of materials, one could reduce the cost of the basic materials by introducing a new philosophy of design. Moreover the constructional advantages of composite construction over fully in situ reinforced concrete are shown to be negated by the extra cost of the materials and also the cost of the shear connectors. These relative advantages and disadvantages of the traditional forms are shown to lead to composite reinforced concrete.

BACKGROUND

Cost of structure

In any comparison of different modes of construction the only comparison which really matters is that of total cost, this being the sum of the cost of the materials, cost of construction and cost of obtaining the required fire resistance.

A discussion of such costs in general terms is best obtained by a comparison of two traditional forms of concrete construction, for example in situ reinforced concrete and normal composite construction (Fig. 1.1). From the material standpoint the reinforced concrete is the cheaper of the two, because the steel content is used much more efficiently, i.e. close to the soffit, giving a near maximum value of the lever arm between the centroids of the compressive and tensile forces. The lever arm in the case of a composite beam at its ultimate load is appreciably less, and hence much more steel is required for the same ultimate resistance moment.

From the constructional standpoint, composite construction is the cheaper. A steel framework of beams and columns can be quickly erected. From this the concrete floors can be constructed much more

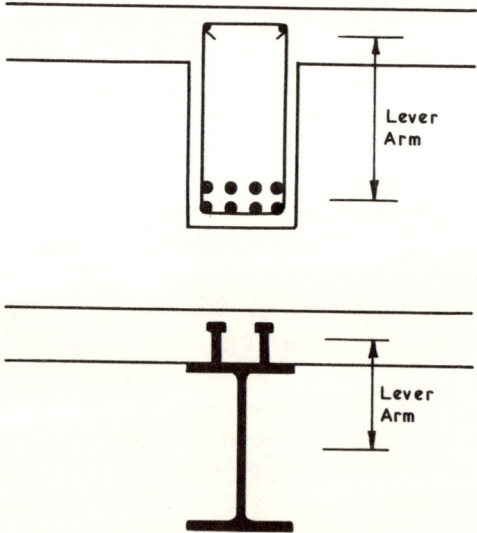

Fig. 1.1. Comparison of two traditional forms of concrete structures

easily than with the completely in situ reinforced concrete structure, particularly when precast concrete units or profiled steel sheeting are used in conjunction with the steel frame to facilitate the construction of the floors. From the fire resistance standpoint reinforced concrete has some advantage, although for multi-storey structures fire protection will often be required for both types. Since both types are currently used for the construction of multi-storey structures it would seem that the total costs are not so different, and therefore that the advantages of the one are more or less counterbalanced by the advantages of the other.

It would seem desirable to obtain a new structural material which combines the material advantage of reinforced concrete with the constructional advantages of composite construction. Composite reinforced concrete is a new structural material which goes some way to achieving this.

Material cost of reinforced concrete

The steel part of reinforced concrete is the most expensive and it is here that we must look for possible reduction in cost of the basic materials. In terms of the stress–strain characteristics there is a wide range of steels available (Fig. 1.2). These extend from the low strength mild steel through high yield steel and very high strength steel to the ultra high strength steel. The latter is currently used only for prestressing concrete, not reinforcing it. Indeed traditional reinforced concrete is confined to the lower strengths of steel, mild steel and high yield steel. Attempts have previously been made to introduce the very high strength steel but, in the face of the limits stipulated by codes of practice these have been unsuccessful.

The restriction to the lower strengths of steel is to ensure that reinforced concrete structures meet the serviceability requirements, i.e. that the widths of cracks and the deflexions of beams remain within acceptable limits. If it were possible to ignore serviceability and design only for ultimate load then much higher strength steels could be used for some types of structure and lower costs would result. For example, an ultra high strength steel having a nominal yield stress of 1500 N/mm^2 is six times as strong as mild steel and only 1/6 of the amount of mild steel would be required. Although the ultra high strength steel is more costly per tonne, there is an appreciable economy overall. A graph of the cost of the reinforcement used in the structure against the strength of

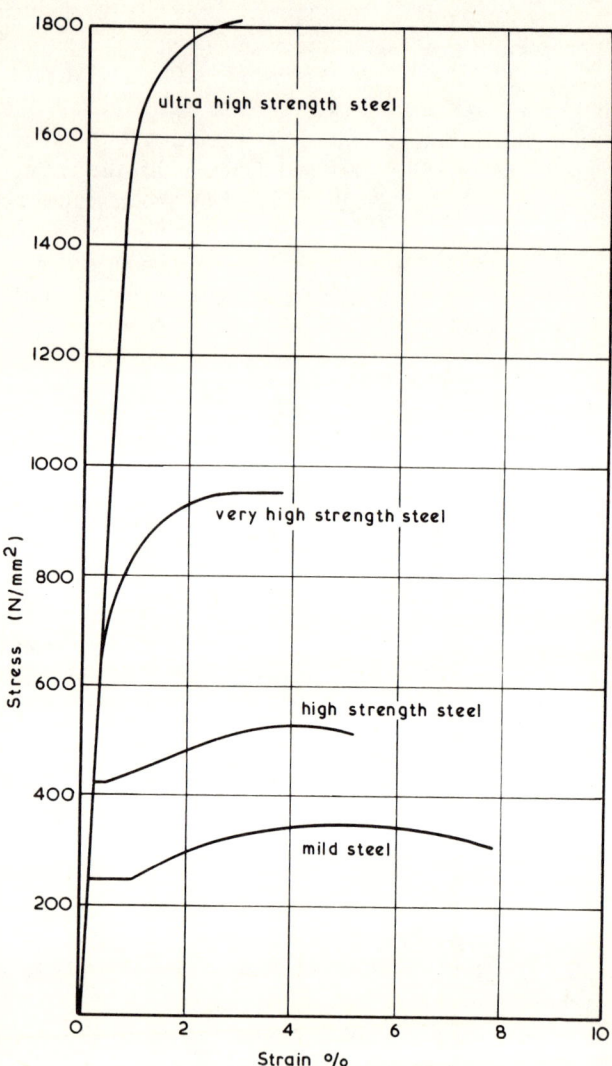

Fig. 1.2. Stress–strain characteristics for steel

the steel would be similar to that shown in Fig. 1.3. However, in order to satisfy serviceability, the design of normal reinforced concrete is restricted to the high end of this graph. Nevertheless it will be shown that for certain types of structure the ultra high strength steels could be used as reinforcement and will satisfy serviceability.

The point is best understood by reference to the load—deflexion characteristics of simply supported T-beams. Consider first a T-beam reinforced with mild steel. Its load—deflexion characteristic would be as in curve 1 in Fig. 1.4. For the most part the characteristic is gently curving as cracking in the web gradually develops under the increasing load. At the onset of yield of the reinforcing bars the deflexions increase much more rapidly and the ultimate load is soon reached.

Consider now a similar beam but reinforced with prestressing strand. For an amount of strand equal to 1/6 of the amount of mild steel the same ultimate load is reached.

This is because for most practical T-beams the neutral axis is within the depth of the slab. The strain diagram in Fig. 1.4 corresponding to the onset of rupture in the compression zone shows that very high strains, and hence very high stresses, are reached at the level of the reinforcement. However this ultimate load now occurs at very high

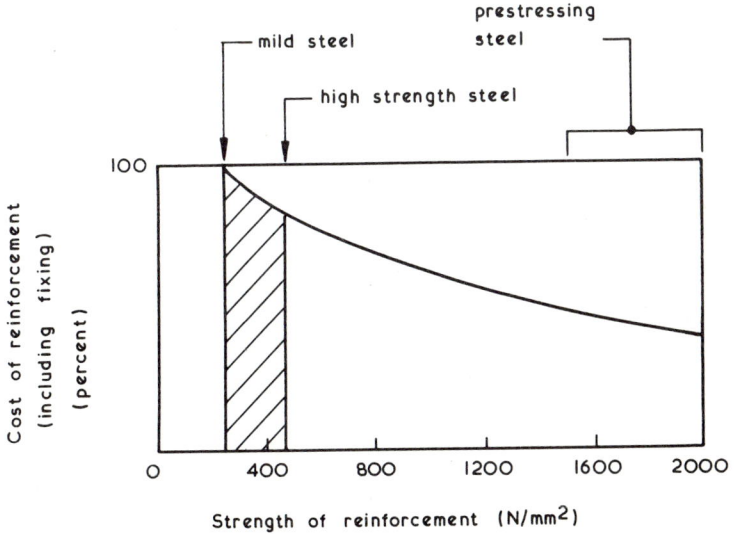

Fig. 1.3. Hypothetical cost curve assuming that serviceability can be disregarded

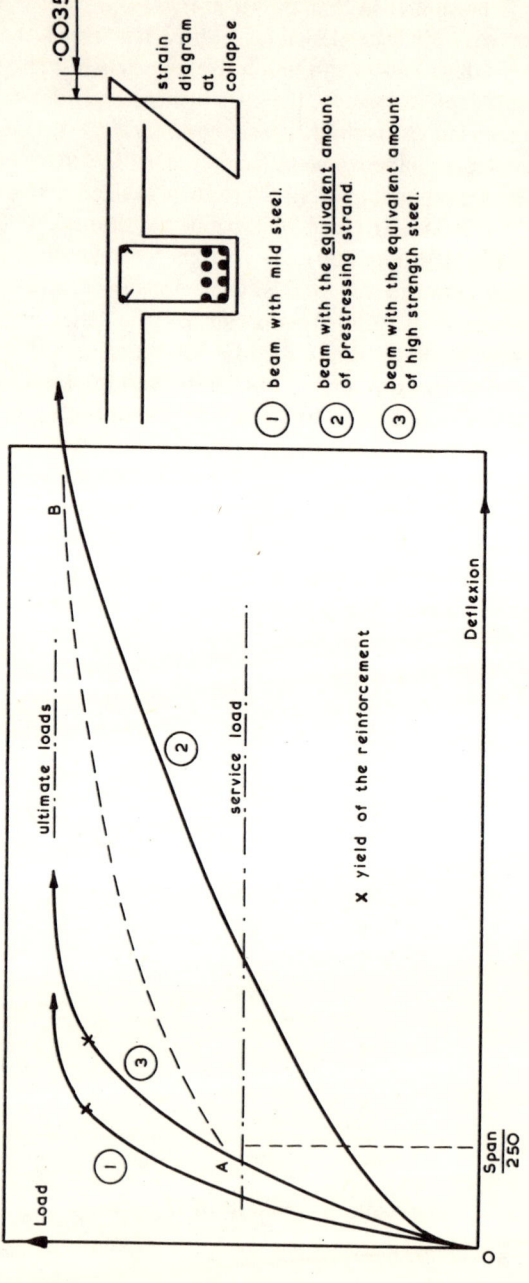

Fig. 1.4. Load–deflexion curves for similar beams that have been differently reinforced

deflexions as indicated by curve 2. Whilst this in itself would not matter the corresponding deflexion of this beam at the service load would be approximately six times the deflexion of the beam with mild steel. The load–deflexion curve corresponding to a similar beam reinforced with the normal high yield steel would be as in curve 3.

The important point to note is that deflexions of beams are only of concern up to the service load. It is here that in design we must ensure that the deflexion satisfies the serviceability criterion of, for example, span/250. It would therefore be quite in order for a beam to have a deflexion characteristic given by OAB; deflexions at the ultimate load are very high but serviceability is satisfied.

Such a curve would be the case for a simply supported T-beam in which the reinforcement is mixed, i.e. partly mild steel and partly pre-stressing strand. For example a beam identical to the previous T-beams but reinforced with half the previous amount of mild steel plus half the previous amount of prestressing strand would have a load–deflexion characteristic such as OAB. At some stage in the loading corresponding to A, the mild steel part of the reinforcement would yield. Beyond this stage the contribution of the mild steel to the resistance moment of the section remains constant, and any additional load is carried only by the increase in stress in the prestressing strand. The rate of deflexion there-fore increases after A. Indeed there is an increasing rate of deflexion as the load increases, due to the region of yielding of the mild steel spreading along the beam. Nevertheless the same ultimate load is reached.

In a beam of such mixed reinforcement the very high strength or ultra high strength steel can be considered as being there to provide the safety margin above the service load – a range of loading which applies only in a laboratory and does not occur in real structures, but which of course must be provided for, just in case.

Thus one aspect of the argument for a new structural material is that it is possible to reinforce concrete using a mixed reinforcement of mild steel with ultra high strength reinforcement in such proportions as to attain the required ultimate resistance moment and to satisfy the serviceability criteria.

Cost of composite construction

In the earlier discussion of total cost, it was stated that composite (steelwork–concrete) structures had the advantage in construction but were extravagant in steel. Another disadvantage is in fact the cost of the shear connectors. Shear connectors are essential to ensure composite action between steelwork and concrete, but their cost is quite significant. For normal composite beams the number of connectors is approximately proportional to the cross-sectional area of the steel section used. It would certainly make composite construction more competitive if the number of the shear connectors could be reduced.

One method of speeding construction in this field is to use some form of unit spanning between the steel beams which allows the in situ concrete to be placed without formwork (Fig. 1.5). Such units can either be precast concrete units or profiled steel sheets. This arrangement not only facilitates construction but improves the efficiency of the section in that the haunch increases the lever arm between the internal compressive and tensile forces. The size of the steel beam is thereby reduced.

In view of the earlier statement regarding the proportionality of cross-sectional area and the number of shear connectors, it might be thought that this method of construction has also reduced the number of shear connectors, but this is not necessarily the case. Unfortunately the strength of the shear connectors in this new situation has been

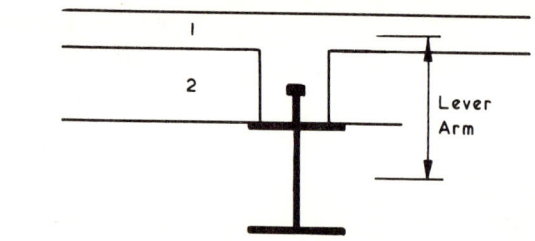

1   in-situ concrete
2   precast concrete

Fig. 1.5. Construction of composite floors using precast concrete

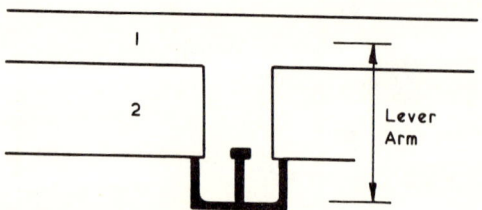

Fig. 1.6. Efficiency of the steel improved by using channels

Separation forces would
cause cracking

Cracking now prevented
by using reinforcement

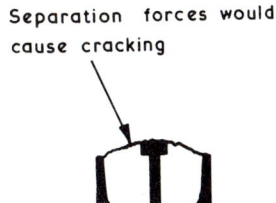

Fig. 1.7. Separation forces cause horizontal cracking of web

Fig. 1.8. Link reinforcement prevents horizontal cracking

reduced, because the concrete surrounding the shear connector is not supported to the same extent as when the connector is at the level of the slab, and as a result bursting of the concrete at the sides of the connector occurs. To allow for this lower strength in design, correspondingly more shear connectors must be provided. Hence, although this procedure has improved the speed of construction and reduced the size of steel section required, the number and cost of shear connectors may have increased.

However, it is possible to go one step further with improvement in this respect: instead of using a steel I-beam, use a steel channel (Fig. 1.6). This is almost as good as the I-beam from the constructional point of view in the sense that pre-cast units can be supported on it and hence the in situ part of the concrete floors can be easily constructed. In doing this the efficiency of the steel is considerably improved. The centroid of the steel channel is very low, quite near the soffit, giving a lever arm which is a high proportion of the full depth. The cross-sectional area of the steel section is thereby reduced. Moreover the strength of the shear connector has been improved compared with the previous deep haunched form of construction, because the concrete at the level of the shear connectors is now confined within the channel and cannot burst outwards. The strength of the shear connector is again determined by its actual shearing strength and not by the bursting of the concrete.

Thus, normal composite construction has been changed to something very close to reinforced concrete — indeed in effect it is reinforced concrete with external reinforcement. However, as shown in Fig. 1.6 it would not work. Although the horizontal forces between the steel channel and the concrete have been provided for, the vertical forces tending to separate the channel and concrete have not, and longitudinal cracking would occur above the shear connectors as shown

1.   mild steel channel.

2.   prestressing strand.

3.   stud shear connectors.

4.   precast units with
        in-situ concrete topping.

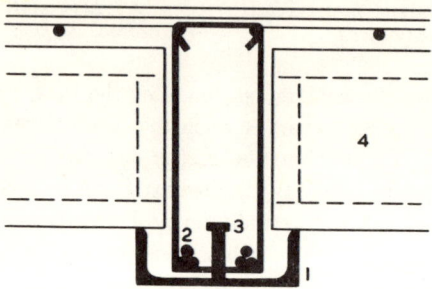

Fig. 1.9. Composite reinforced concrete

in Fig. 1.7. It would not be economic to provide very long shear connectors to carry these separation forces into the compression zone. Instead longitudinal and transverse reinforcement is used as in Fig. 1.8. Any tendency to separate now produces compressive struts in the concrete between the underside of the stud head and the longitudinal bars, the latter conveying these forces to the links.

The links are not just for this purpose, however. Such links have to be provided in normal reinforced concrete T-beams to resist the shear forces, and the same shear links will also resist the separation forces. With regard to the internal longitudinal steel, this can be used in conjunction with the channel to resist the applied bending moment. The longitudinal reinforcement within the channel could be the ultra high strength steel, i.e. prestressing strand. This arrangement of mixed steel forms the basis of the proposals for a new structural material.

COMPOSITE REINFORCED CONCRETE

The arrangement summarized in Fig. 1.9 is the proposed new form of construction — composite reinforced concrete. It is essentially reinforced concrete in which the reinforcement consists of a mild steel channel forming the soffit of the beam and acting in conjunction with prestressing strand to resist bending moments.

Method of construction

The improvement in the speed of construction stems from the use of the mild steel content of the beam to support precast concrete units or similar spanning between the beams. It is anticipated that the proposal for mixed steel could be extended to columns by using battened steelwork columns formed from angles or channels in conjunction with internal reinforced concrete. Fig. 1.10 illustrates the type of construction which could then result. A steel framework of vertical and horizontal members would first be erected followed by the precast concrete units. In situ concrete for the floors and columns would tie the whole into a rigid-jointed framework. In connection with this suggestion the following points should be noted.

(a) No site connections between the columns and the horizontal channels would be required, the channels being simply supported on the cross-battens of the columns. (This would, however, limit the erection of the bare steel framework to one or two floors above a completed floor to avoid instability during erection.)

(b) The reinforcement cage for the beam could be fitted in the channel at ground level prior to lifting the channel into position.

(c) Transverse bolts would be required to retain the reinforcement cage within the channel during lifting, these bolts also acting as shear connectors and as supports for brackets for a false ceiling or services below the beam.

(d) The battened column using angles or channels is technically a very efficient column and can be designed to have no weak axis.

One of the disadvantages of the proposed system is the greater flexibility of the channel compared with the steel I-beam of normal composite construction. This will necessitate propping the channel during construction, whereas in normal composite construction this can be avoided, although propping of the I-beam is also often stipulated by the designer in order that the steel can be used more efficiently. Propping of the channels at one-third points along the beam will normally be sufficient. For the construction of multi-storey buildings this should not be too onerous.

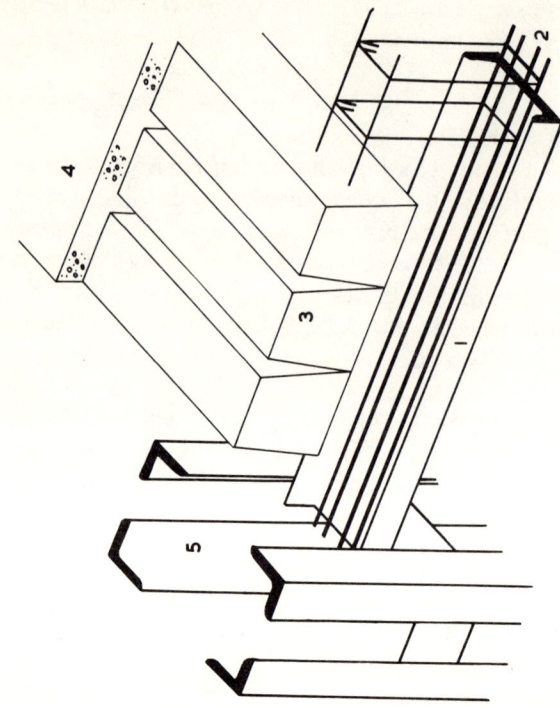

Perspective view showing ease of assembly.

1. Structural steel channel.
2. Very high strength reinforcement.
3. Prefabricated units.
4. In situ concrete.
5. 'Mini' steel column using battened angles.

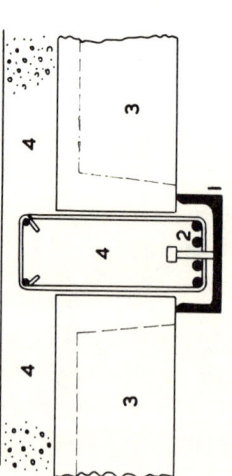

Section through beam

Fig. 1.10. Possible method of construction

Architectural points

Some features of the new method of construction are that

(a) the arrises of the soffit of the beam will always remain sharp and cannot be damaged during construction

(b) there is a complete absence of cracking at the soffit of the beam

(c) only a very small depth of construction beneath the transverse precast units is required; this is in contrast to composite construction

(d) in comparison with normal reinforced concrete the beams of composite reinforced concrete can be made much more compact.

Items (a) and (b) would not often be advantageous because false ceilings will usually be required anyway. Items (c) and (d) are considered to be important advantages.

Comparisons

In the construction of multi-storey buildings composite reinforced concrete should be more economical in total cost than normal in situ reinforced concrete since

(a) prestressing strand has replaced an equivalent amount of mild or high strength steel

(b) the mild steel content of the beam is used in a form to facilitate construction.

It should be more economical than normal composite construction because

(a) the cost of the steel is appreciably lower

(b) the number of shear connectors is reduced considerably

(c) the need for web stiffeners is eliminated

(d) most of the constructional advantages are retained.

It should be more economical than prestressed concrete because the prestressing operation is prohibitively costly for in situ multi-storey framed structures.

EXPERIMENTAL DATA FOR FLEXURE

Many simply supported beams in composite reinforced concrete have been tested to verify that the material is satisfactory from the technical

Table 1.1. Flexural tests on beams with Bristrand by Burdon[1]

Beam	Load causing flexural failure			Theoretical stress in strand at failure: N/mm²	Maximum stress in channel at the service load		
	Theoretical P_u:kN	Experimental P_e:kN	Ratio P_e/P_u		Theoretical f:N/mm²	Experimental f_e:N/mm²	f_e/f
4D1	167	164	0·98	950	255	275	1·08
4D2	205	208	1·01	850	280	270	0·96
6D1	310	325	1·05	860	260	260	1·00
6D3	256	267	1·04	850	235	225	0·96
6S1	156	166	1·06	770	240	240	1·00
6S2	159	176	1·11	700	240	230	0·96

Table 1.2. *Flexural tests on beams with prestressing strand by Rankin[3] and Halliday[4]*

Beam	Theoretical depth of compression zone at failure: x		Load causing flexural failure			Stress in strand at failure		Maximum stress in channel at the service load		
	mm	x/D	Theoretical, P_u:kN	Experimental, P_e:kN	Ratio, P_e/P_u	Theoretical: N/mm²	Experimental: N/mm²	Theoretical, f:N/mm²	Experimental, f_e:N/mm²	f_e/f
P1	94	0·31	188	193	1·03	1310	1400	303	280	0·92
P2	66	0·19	217	227	1·05	1690	1570	241	230	0·95
P3	96	0·27	279	283	1·01	1450	1580	305	300	0·98
P4	95	0·29	245	267	1·09	1330	1300	255	255	1·0
H1	85	0·24	314	302	0·96	1640	—	304	366	1·2
H2	100	0·28	295	286	0·97	1525	—	298	332	1·1

Notes
1. The theoretical values corresponding to the ultimate load have been calculated using the strain compatibility method given in CP 110, but ignoring partial safety factors.
2. The experimental values of stress in strand at failure have been estimated by extrapolating from the last measured strains. (As a result they are different to the values given in reference 3, those being calculated from the last measured strains.)
3. The stresses at the service load have been calculated assuming a linear stress–strain relationship; in some beams the calculated value slightly exceeded the yield stress of 300 N/mm².

standpoint. However not all these beams were constructed using prestressing strand for the internal reinforcement, which is the type of reinforcement now advocated. Until comparatively recently, the characteristic strengths of the prestressing strand were thought to be too high to be achieved in beams which are essentially a form of reinforced concrete. As a result many of the earlier tests were carried out using Bristrand, a very high strength 3 wire strand having a nominal yield stress of 700 N/mm² and an ultimate stress of approximately 950 N/mm². The tests using Bristrand carried out by Burdon have been described elsewhere,[1] but for convenience the main results are summarized in Table 1.1.

The main conclusions from these tests were that

(a) the ultimate resistance moments of the beams could be estimated with good accuracy by the strain compatibility method of calculation given in CP 110[2]

(b) the maximum stresses attained in the reinforcement at the ultimate load were approximately twice the limit stipulated in CP 110 and yet the beams were completely satisfactory at the service load

(c) the stresses in the channel at the service load could be estimated quite accurately by the elastic method of analysis used for normal reinforced concrete.

Since that time market forces have prevented commercial production of Bristrand. As a result it has become clear that, if the idea of composite reinforced concrete is to be successful, only very high strength steel already in commercial use would be economically viable. It was at that time that attention was turned to prestressing strand and further tests were made to ascertain that the general results from the tests using Bristrand were still applicable for beams reinforced with prestressing strand. Tests by Rankin have been described fully elsewhere,[3] but the main results from his beams reinforced with prestressing strand and failing in flexure, together with two beams tested by Halliday,[4] are summarized in Table 1.2.

The results of the tests using prestressing strand indicated that the previous general conclusions still apply despite maximum stresses of nearly 1600 N/mm² (3½ times the maximum stress recommended in CP 110) being attained in the strand reinforcement at the ultimate load.

Regarding the quoted value of 1600 N/mm², the experimental values of stress in the prestressing strand were estimated from measured strains by extrapolating from the last measured values. Inevitably, then, they are approximate.

The values quoted are also different to the values given in reference 3 where the stresses were calculated from the last measured strains. Since there was sometimes an appreciable difference between the last load at which strains were measured and the load causing failure it was felt that the maximum stresses quoted in reference 3 gave a misleading impression. Although extrapolating from the last measured strains introduces some error, it is felt that these values are nearer the true ones.

Only in one of the laboratory beams was the ratio of the depth of the compression zone to overall depth less than 1/5. As a result the theoretical stresses in the strand at failure in all beams except this one were less than the full characteristic strength of the strand. In practical beam—slab structures the effective widths of the flanges of the T-beams will be appreciably wider than the flanges of the laboratory beams. This will lead to relatively shallow compression zones and therefore to the full characteristic strength of the prestressing strand being theoretically possible to attain. It would seem desirable that tests typical of such practical structures be carried out prior to the use of composite reinforced concrete in practice.

CALCULATION OF ULTIMATE RESISTANCE MOMENT

Method for the general case

The tests have shown that the strain compatibility method for the calculation of the ultimate resistance moment given in CP 110 for reinforced concrete can also be applied to beams of composite reinforced concrete. The procedure, which is illustrated in Fig. 1.11, can be summarized as follows.

(a) Assume a depth of compression zone at flexural failure of the beam. (In T-beams of practical proportions the neutral axis will normally lie within the slab.)

(b) Draw the strain diagram corresponding to the ultimate load using the assumed depth of compression zone. (The maximum strain in the outer fibres of the concrete is taken as 0.0035.)

(c) From the strain diagram determine the strains at the centroids of area of the channel and the strand.

(d) From the stress—strain characteristics of the two steels obtain

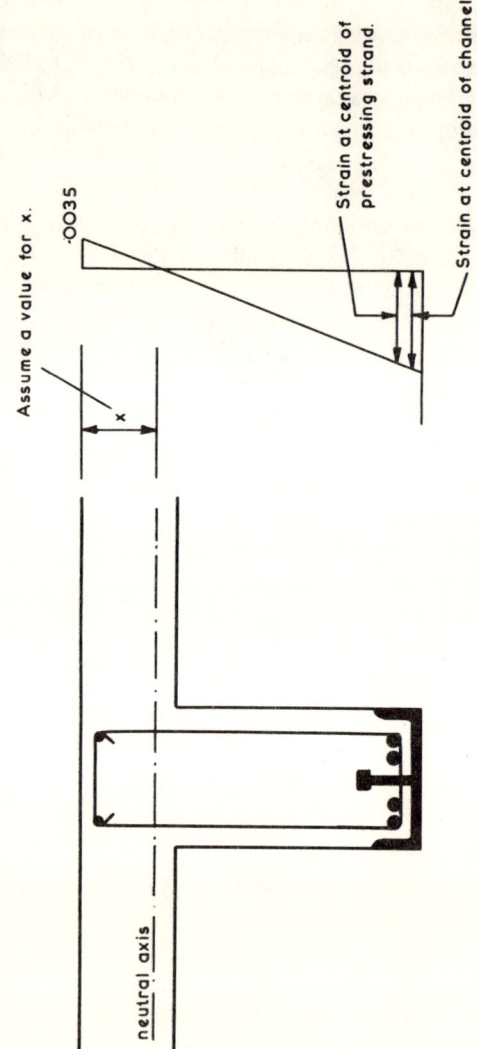

Fig. 1.11. Procedure for calculating the ultimate resistance moment

the stresses corresponding to these strains and hence the forces in the channel, F_{ch}, and the strand, F_{st}. (In the case of design the design stress–strain curve in CP 110 is used for the strand reinforcement.)

(e) Calculate the force in the compression zone of concrete, F_c, assuming an average stress of $0.6 f_{cu}$ in the case of laboratory beams or $0.4 f_{cu}$ in the case of the design of beams for practical use.

(f) If $F_{ch} + F_{st} = F_c$ then the assumed value of the depth of the compression zone was correct.

(g) Determine the lever arm between the centroid of the channel and the centroid of compression, z_{ch}, and the lever arm between the centroid of the strand and the centroid of compression, z_{st}.

(h) The ultimate resistance moment of the beam is then given by

$$M_{ult} = F_{ch}z_{ch} + F_{st}z_{st}$$

(i) The final step is to ascertain that at the bending moment corresponding to the service load the maximum stress in the channel does not reach the yield stress of the steel.

Usual method for the design case

For beam–slab structures of practical proportions the depth of the compression zone is such that the mild steel of the channel reaches its yield stress and the strand reinforcement reaches its characteristic strength. Hence for design

$$F_{ch} = \frac{A_{ch}f_y}{1.07} = 0.93 f_y A_{ch}$$

where 1.07 is the material safety factor recommended in the draft code of practice for the use of composite construction in buildings[5] and f_y is the characteristic yield stress of the mild steel channel, currently 250 N/mm^2,

$$F_{st} = \frac{P_{st}}{1.15} = 0.87 P_{st}$$

where P_{st} is the characteristic strength of the prestressing strand given in Table 30 of CP 110, and 1.15 is the materials safety factor recommended by CP 110.

This direct calculation of the tensile force at failure enables the depth of the compression zone to be calculated directly, and hence also the distances of the lever arms.

It will occasionally be necessary to check from the strain diagram that the strain corresponding to the characteristic strength of the strand can be attained. The approximate required strain is given by Fig. 3 of CP 110 as

$$0.005 + \frac{1630}{1.15 \times 200 \times 10^3} = 0.012$$

To attain this strain at the level of the strand would require that the depth of the compression zone at failure is not more than approximately 1/5 of the overall depth of the beam. Thus where this is the case no check on strain would be required.

Ratio of design forces at ultimate load

In order to facilitate design a guide to the relative forces in the channel and the strand at the ultimate load is desirable. Such a guide can be achieved by a series of approximations.

If it is assumed that the lever arms of the forces in the channel and strand are equal, we can write

$$M_{ult} = (0.93 f_y A_{ch} + 0.87 f_{st} A_{st})z$$

Since

$$\frac{f_{st}}{f_y} = \frac{1630}{250} = 6.5$$

then

$$M_{ult} = (0.93 f_y A_{ch} + 5.7 f_y A_{st})z \tag{1}$$

If it is further assumed that the lever arms of the forces at the bending moment corresponding to the service load can be taken as the same value z, then

$$M_{sl} = (A_{ch} + A_{st})fz$$

where f is the stress at the level of the centroid of steel at the service load.

5.   Channel   254 x 89 mm x 36 Kg/m

6.   4 No. 15mm prestressing strand

7.   19 x 100 mm studs (40 No.)

8.   Shear reinforcement (8 mm)

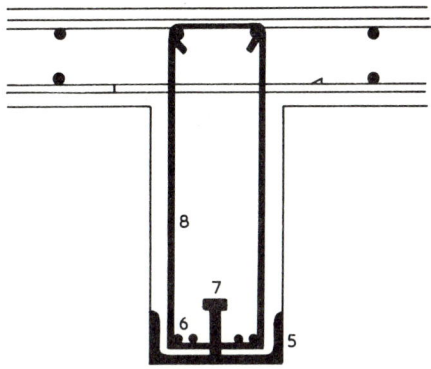

Fig. 1.12. Section of beam used for illustrating the calculation of the ultimate resistance moment

Since at the service load the stress in the bottom fibres must not exceed 250 N/mm², assume that

$$f = 0.95\, f_y$$

Thus

$$M_{sl} = (A_{ch} + A_{st})0.95\, f_y z \tag{2}$$

For a ratio of ultimate bending moment and service load given by

$$M_{ult} = 1.5\, M_{sl}$$

equations (1) and (2) can be combined to give

$$0.93\, f_y A_{ch} + 5.7\, f_y A_{st} = 1.5(A_{ch} + A_{st})0.95\, f_y$$

leading to

$$A_{ch} = 8.7\, A_{st}$$

From

$$F_{ch} = 0.93\, f_y A_{ch}$$

and

$$F_{st} = 0.87 f_{st} A_{st}$$

we obtain

$$F_{st} = 0.7 F_{ch}$$

Thus at the ultimate load the design force in the strand will be approximately 0.7 times the yield force of the channel. This figure is approximate only and is to be used to facilitate the proportioning of the first trial section in design. In the finally adopted section the actual ratio of the two forces will be slightly different to this ratio as the above approximations will be slightly in error.

Example

The design ultimate resistance moment of the section shown in Fig. 1.12 will be calculated. The span of the beam is assumed to be 9 m with the reinforced concrete slab spanning transversely 6 m between beams.

Area of steel channel	$= 4552 \text{ mm}^2$
Design yield force in channel $= 0.93 \times 0.25 \times 4552$	$= 1060 \text{ kN}$
Design ultimate force in strand $= 4 \times 0.87 \times 227$	$= 790 \text{ kN}$
Total ultimate tensile force	$= 1850 \text{ kN}$
Effective width of slab $= 9/5 + 0.25$	$= 2.05 \text{ m}$
Depth of compression zone $= 1850/(0.4 \times 30 \times 2.05)$	$= 75 \text{ mm}$
Lever arm from centroid of channel $= 640 - 24 - 38$	$= 578 \text{ mm}$
Resistance moment provided by channel $= 1060 \times 0.578$	$= 612 \text{ kNm}$
Lever arm from centroid of strand $= 640 - 40 - 38$	$= 562 \text{ mm}$
Resistance moment provided by strand $= 790 \times 0.562$	$= 444 \text{ kNm}$
Ultimate resistance moment of beam	$= 1056 \text{ kN}$

As the depth of the compression zone is less than 1/5 of the overall depth there is no need to check the strain at the level of the strand. Nevertheless the calculation is given for purposes of illustration.

Strain at level of strand $= 0.0035 \times (562 - 38)/75$ $\qquad = 0.024$

This strain is much more than is necessary to enable the characteristic strength to be attained and indicates that a more compact section could have been adopted.

Assuming a service load bending moment of 615 kNm, the maximum stress in the channel is calculated as follows.

Approximate lever arm at service load	= 576 mm
Total steel force = 615/0·576	= 1070 kN
Total area of steel = 4552 + 4 × 139	= 5108 mm^2
Mean stress in steel = $(1070 \times 10^3)/5108$	= 210 N/mm^2
Maximum stress in channel	
= 210 × (576 − 38 + 26)/(576 − 38)	= 220 N/mm^2

This is less than the yield stress of the mild steel channel of 250 N/mm^2.

COMPARATIVE DESIGNS

To facilitate the comparison of composite reinforced concrete with the traditional materials of reinforced concrete and composite construction, some designs in each of the three materials will be given.

The structure used for the comparison is a floor of the multi-storey building shown in plan in Fig. 1.13. A reinforced concrete slab forms the floor and spans 6 m between the beams and is continuous over them. The beams span the 9 m width of the building and are assumed to be simply supported. It is the design of a typical beam which forms the basis of the comparison. An imposed load of 4 kN/m^2 is assumed together with a dead load due to finishes and partitions of 1·5 kN/m^2. Two sets of designs have been prepared.

Designs A

The first set of designs aims at a simple comparison of material quantities. For this the same slab and the same depth of beam have been

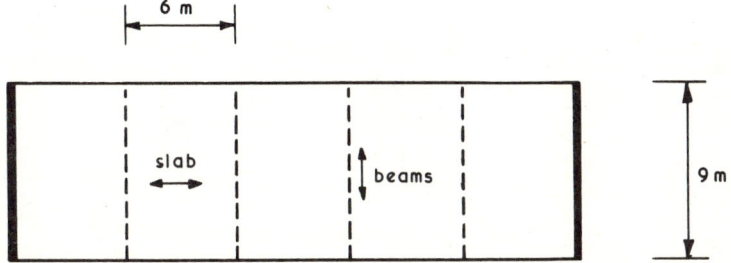

Fig. 1.13. Floor plan used for comparative designs of beams

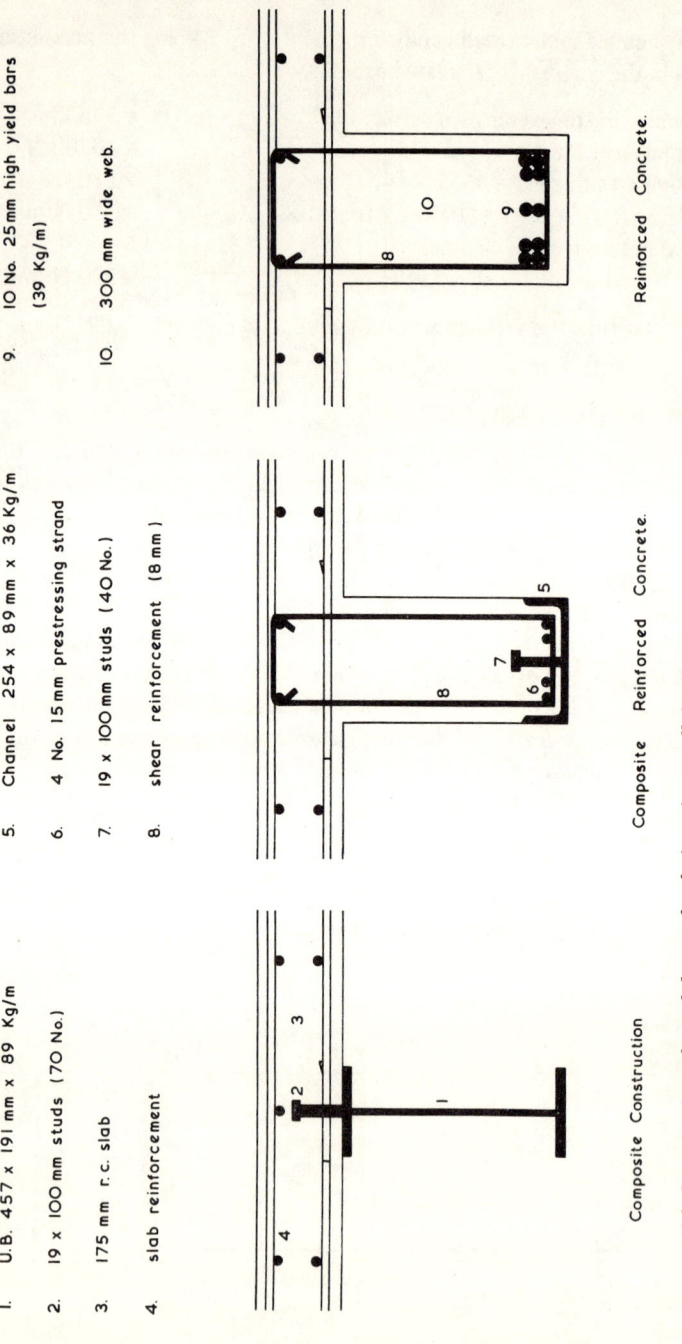

1. U.B. 457 x 191 mm x 89 Kg/m

2. 19 x 100 mm studs (70 No.)

3. 175 mm r.c. slab

4. slab reinforcement

5. Channel 254 x 89 mm x 36 Kg/m

6. 4 No. 15 mm prestressing strand

7. 19 x 100 mm studs (40 No.)

8. shear reinforcement (8 mm)

9. 10 No. 25 mm high yield bars (39 Kg/m)

10. 300 mm wide web.

Composite Construction   Composite   Reinforced   Concrete.   Reinforced   Concrete.

Fig. 1.14. Cross-sections through beams for designs A; overall depth 640 mm

1. U.C. 305 x 305 mm x 118 Kg/m

2. 19 x 100 mm studs   (92 No.)

3. troughed steel sheeting acting
    as formwork and bottom
    reinforcement.

4. slab reinforcement.

5. Channel 305 x 102 mm x 46 Kg/m

6. 4 No. 15 mm and 2 No. 12 mm
    prestressing strand (6 Kg/m)

7. 19 x 100 mm studs (52 No.)

8. precast units 325 mm deep
    with 75 mm in-situ topping

9. 15 No. 25 mm high yield bars
    (58 Kg/m)

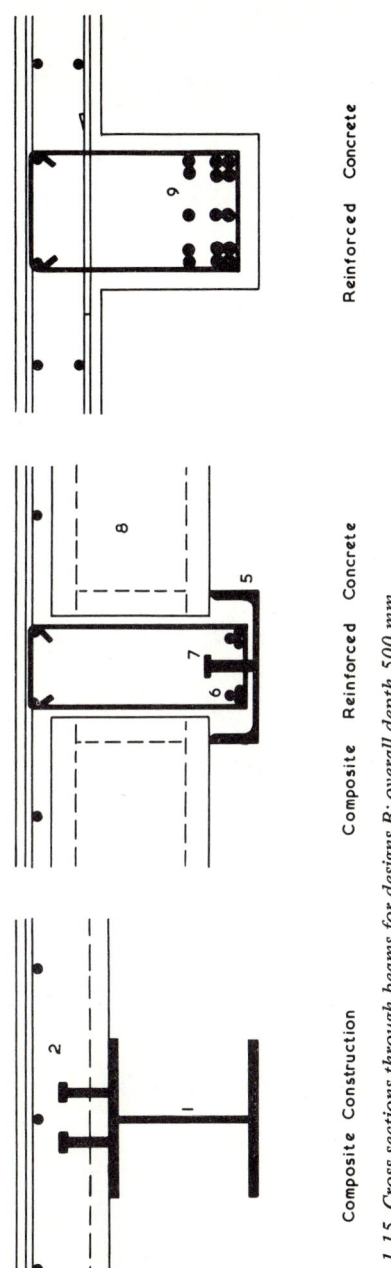

Composite Construction   Composite Reinforced Concrete   Reinforced Concrete

Fig. 1.15. Cross-sections through beams for designs B; overall depth 500 mm

Table 1.3. Quantities and costs* of materials in beams of designs A

Structural material	Steelwork		Stud shear connectors		Reinforcement†		Prestressing strand		Total cost: £
	Quantity: kg	Cost: £	Quantity	Cost: £	Quantity: kg	Cost: £	Quantity: kg	Cost: £	
Composite construction	800	148	70	16	–	–	–	–	164
Reinforced concrete	–	–	–	–	403	72	–	–	72
Composite reinforced concrete	320	59	42	9	48	8	39	10	86

*The rates used for the calculation of costs are those for the materials only and do not include fixing or fabrication and erection.
†The reinforcement quantities include the shear reinforcement which for the reinforced concrete beam amounts to 53 kg.

used in the three designs and there has been no attempt to incorporate special features which might in practice be used. The depth of the beams was largely determined by the composite construction design, this depth being the minimum possible for these particular beams in normal composite construction using universal beam sections.

The cross-sections of the beams in the three materials are shown in Fig. 1.14, the calculations being summarized in Appendix 1. The design of the beam in composite construction was in accordance with the recommendations of the new draft code of practice whilst the reinforced concrete design was in accordance with CP 110. The design in composite reinforced concrete was in accordance with the appropriate recommendations of both these codes.

The quantities of materials required for the three designs are summarized in Table 1.3, and these have been translated into costs using 1978 rates. The costs do not include fixing the reinforcement or fabrication and erection of the steelwork.

As might be expected from such a comparison the total material cost of the beam of composite construction is higher than those for reinforced concrete and composite reinforced concrete. This stems from the fact that for both reinforced concrete and composite reinforced concrete the steel for resisting the bending stresses is used very efficiently near the soffit of the beam, whilst much of the steelwork section of the beam of composite construction is used very inefficiently. For example the lever arm between the centroid of the compression force in the concrete and the tensile force from the top flange of the steel beam is very small.

However, the cost of construction in composite construction would be appreciably lower than that in reinforced concrete. The extra formwork for the beams, together with the slowness of construction of the completely in situ structure, places reinforced concrete at a significant disadvantage regarding constructional costs. Moreover, as previously explained, composite construction allows for the possibility of improving the rate of construction by incorporating precast units or profiled steel sheeting.

The beams of composite reinforced concrete shown in this particular design would be at a similar constructional disadvantage to the beams of normal reinforced concrete. However it will be recalled that composite reinforced concrete has been developed as a form of reinforced concrete with the particular facility of improving the rate of construction, again by utilizing precast units. Moreover the compactness of the

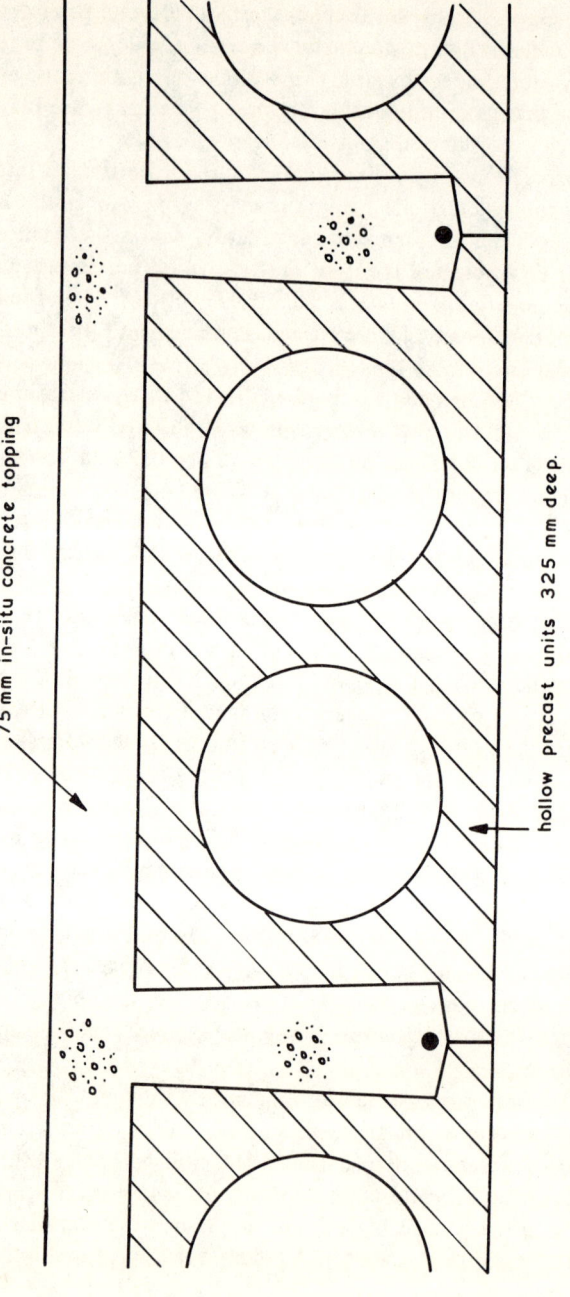

75 mm in-situ concrete topping

hollow precast units 325 mm deep.

Fig. 1.16. Section through slab assumed for design B in composite reinforced concrete

steel in composite reinforced concrete enables the overall depth of the beam to be reduced to that which would not be acceptable in reinforced concrete because of the resulting congestion of reinforcing bars. To illustrate these points a second set of designs has been carried out.

Designs B

The second set of designs is shown in Fig. 1.15 with a summary of the calculations given in Appendix 1. Again the three forms have the same overall depth, this being reduced compared with designs A.

The composite reinforced concrete design has this time incorporated precast units which form the permanent shuttering for the in situ concrete placed above and between the units (see Fig. 1.16). The units themselves sit on the edges of the flanges of the channel. There is no doubt that this arrangement would greatly facilitate construction compared with the design A in composite reinforced concrete, albeit at the expense of providing the precast units and the extra dead load of the slab. The extra depth of slab does, of course, significantly reduce the amount of the slab reinforcement.

The effect of the reduced depth on the in situ reinforced concrete design results in more longitudinal and transverse reinforcement. In comparison with composite reinforced concrete, the reinforced concrete design becomes relatively inefficient for the reduced depth, the effective depth being now significantly less than the overall depth of the beam. The resulting amount of longitudinal reinforcement for this depth of beam would seem to be unacceptable, giving a congestion of bars and high fixing costs. The beam in composite reinforced concrete would appear to be quite acceptable from this standpoint.

The design in composite construction would prove economically unacceptable in practice as the reduced depth now requires a very heavy universal column section. This design shows a method of constructing the floor slab using profiled steel sheets as permanent shuttering. In the designs for composite construction propping of the steel beam during construction was assumed. It would be possible to eliminate propping altogether, but only at the expense of a larger steel section.

In the case of composite reinforced concrete propping could not be eliminated and would always be required. Indeed, whereas propping of the I-beam in composite construction would only be necessary at mid-span, the channel in this particular case would need to be propped at the quarter span points in addition to that at midspan. Nevertheless such propping should not prove too onerous.

From this particular set of designs it would seem that composite reinforced concrete offers the only possible mode of construction, this stemming from the compactness of the steel used to provide the tensile force of the resistance moment.

SUMMARY

Composite reinforced concrete appears to be most advantageous for the construction of heavily loaded multi-storey buildings where it is desired to use precast flooring units in conjunction with an in situ frame for which beams of minimum depth are required. No temporary formwork for the concrete is needed and the depth of beams required beneath the floor units is very small.

For simplicity the comparisons made here have been confined to beams with sagging bending moments only. Composite reinforced concrete does, in fact, have advantages in resisting hogging bending moments also, and this is dealt with in chapter 2.

REFERENCES

1. TAYLOR R. and BURDON P. Tests on a new form of composite construction. *Proc. Instn Civ. Engrs,* Part 2, 1972, Vol. 53, Dec., 471–485.
2. BRITISH STANDARDS INSTITUTION. 'CP110: Structural use of concrete.' British Standards Institution, London, 1972.
3. TAYLOR R., MILLS P. E. and RANKIN R. I. Tests on concrete beams with mixed types of reinforcement. *Mag. Concr. Res.,* 1978, Vol. 30, June, No. 103.
4. HALLIDAY K. D. 'Tests on composite reinforced concrete beams with bar shear connectors.' University of Manchester internal report, 1978.
5. BRITISH STANDARDS INSTITUTION. 'Draft Code: Use of structural steel in buildings, Part 3, Composite construction.' British Standards Institution, London, Aug., 1976.

CHAPTER 2

Problems of continuity

The flexural behaviour of composite reinforced concrete discussed in chapter 1 was related to sagging bending moments only. The regions of hogging bending moments in continuous beams and rigid-jointed frames introduce new aspects of behaviour. The channels forming the soffit to the beams will be continuous from column to column and, where appropriate, continuous through the column. Thus in the region of the columns the steel channels will be in compression, as will also the concrete contained within the channel.

The strand reinforcement will not necessarily be continuous from column to column. Some of the strands would be stopped-off between the columns and the points of contraflexure, leaving probably only two strands to continue to the column on which to position the shear links and to effect the necessary transfer of forces discussed previously.

On the tension side of the beam the tensile force is taken by normal reinforcing bars. (The possibility exists of bending-up the strands from the bottom of the beam, where they are no longer required for sagging bending moments, to help the top reinforcing bars to resist the hogging bending moments, but this would probably not be economic.)

Regarding this arrangement for resisting hogging bending moments the following points should be noted.

(a) The channel is a very suitable shape for resisting compressive forces and is able to attain its full yield capacity without prior buckling of the web and flanges.

(b) The strand is an unsuitable shape for resisting compression. Although its confinement by the surrounding concrete would enhance its compressive strength, its small cross-sectional area at the support makes it possible to ignore its contribution to the total compressive force.

(c) The concrete contained within the channel will be confined by the channel and this will enhance its strength to some extent.

(d) In addition to the strength of the concrete being enhanced by the confinement, so also is its strain capacity. This in turn considerably enhances the rotational capacity of hinge regions beyond that normally associated with concrete structures.

Composite reinforced concrete has, therefore, several advantages over the traditional materials in resisting hogging bending moments.

(a) In the design of composite construction a check needs to be made on whether buckling of the flanges of the I-beam would occur prior to yielding in compression; in the design of composite reinforced concrete no such check is necessary as all the channel sections would yield fully before any buckling occurred.

(b) In steelwork and composite construction the connections necessary to effect continuity are costly, and as a result are avoided where possible in favour of simple connections and simply supported beams; the possibility with composite rein-

I.   Pair of steel channels positioned with the
      tips of flanges touching.

2.   Annulus between channels filled with concrete.

3.   Spacing of links constitute one of the variables.

4.   Steel platens of the testing machine.

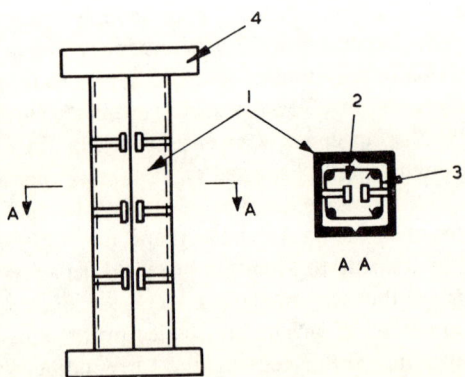

Fig. 2.1. Tests to determine the strength of concrete contained by the channels

forced concrete of obtaining channels continuous through the column (e.g. the type of column in Fig. 1.10) would seem to offer a cheaper means of effecting continuity in multi-storey structures.

(c) For reinforced concrete plastic design can only be used in conjunction with checks on the rotational capacities of sections, by reference to the depth of the compression zones, making for tedious trial and error procedures; in composite reinforced concrete the rotational capacities are likely to be such that no checks are necessary. (This point is discussed in more detail on p. 53).

COMPRESSIVE STRENGTH OF CONCRETE CONTAINED BY CHANNEL

Numerous examples occur whereby the strength of concrete is enhanced by forces which restrain its lateral expansion. The cube test itself is one such example, the steel platens of the testing machine providing the restraining forces to give strengths approximately 30% higher than the strength of concrete where such restraint is absent. A similar type of restraint occurs in a concrete hinge where the strength of the throat can be several times the cube strength of the concrete. In order to determine the compressive strength of the concrete contained by the channel in composite reinforced concrete it is necessary to carry out tests which simulate the same conditions.

Test details

Tests carried out in the University of Manchester are shown diagrammatically in Fig. 2.1. The specimens consist of two channels with ends machined to give identical lengths. These are positioned with the tips of their flanges together. The annulus so formed is filled with concrete which is reinforced longitudinally and transversely with amounts of strand and shear links considered typical of beams. Stud shear connectors provided the same type of connection between channel and concrete as in beams. Such specimens were tested in standard compression machines.

The concrete is restrained differently in the two directions: in one by the channels and in the other by the links. The degree of restraint offered by a channel will depend on the stage of loading. For example, since the longitudinal strains in the steel channel and the concrete are constrained to be equal, then restraint to lateral expansion in the early

Table 2.1. Results of tests on compression specimens

Size of channel: mm × mm	Strength of specimen without concrete: kN	Designation of specimen	No. of reinforcing links	Diameter of reinforcing links: mm	Ultimate load of specimen: kN	Increase in carrying capacity due to concrete: kN	Apparent strength of the concrete		Mean measured longitudinal strain × 10³	
							f: N/mm²	Ratio: f/f_{cu}	At or near ultimate load	At 90% of ultimate load
102 × 51	1010	4–0A	–	–	1350	325	42	1·05	4	2
		4–0B			1320				2	2
		4–6A	6	5	1400	410	53	1·2	9	2
		4–6B			1440				6	3
		4–9A	9	5	1410	435	56	1·4	5	3
		4–9B			1480				7	4
		4–12A	12	5	1400	420	54	1·35	6	3
		4–12B			1460				4	2
127 × 64	1400	5–0A	–	–	1710	320	26	0·65	9	2
		5–0B			1730				11	2
		5–6A	6	5	1985	572	46	1·00	8	3
		5–6B			1960				8	3
		5–9A	9	5	1710	380	30	0·75	11	3
		5–9B			1850				8	3
		5–12A	12	5	1930	495	40	1·0	12	6
		5–12B			1860				10	5
152 × 76	1470	6A–9	9	5	2290	820	44	1·1	12	4
		6A–12	12	5	2130	660	35	0·9	10	5
152 × 89	1890	6B–9	9	5	2780	890	42	1·05	12	6
178 × 89	2080	7B–7	7	5	2950	870	35	0·9	9	4
		7B–9	9	5	2915	835	34	0·85	9	4
		7B–14	14	5	2825	745	30	0·75	9	4

stage of loading will be small or nil. In the later stage of the test when the concrete is at incipient failure and attempting to fail by lateral expansion, then the effect of the confinement by the channel would be appreciable.

The restraint to the concrete offered by the links would operate through all stages of loading, but clearly at the critical stage of incipient failure of the concrete the restraint offered by the links in the one direction will be much less than the restraint by the channel in the other direction.

Tests have shown that the degree of enhancement of strength of triaxially restrained concrete is dependent largely on the weakest restraint. Thus for the situation in composite reinforced concrete the degree of enhancement is determined by the links and therefore high degrees of enhancement of strength of the contained concrete cannot be expected.

However, there is one very important effect that the channels will have on the effective strength of the concrete. The concrete outside the shear links will be brought within the restraining influence of the links by the fact that the channels and links are connected by the stud shear

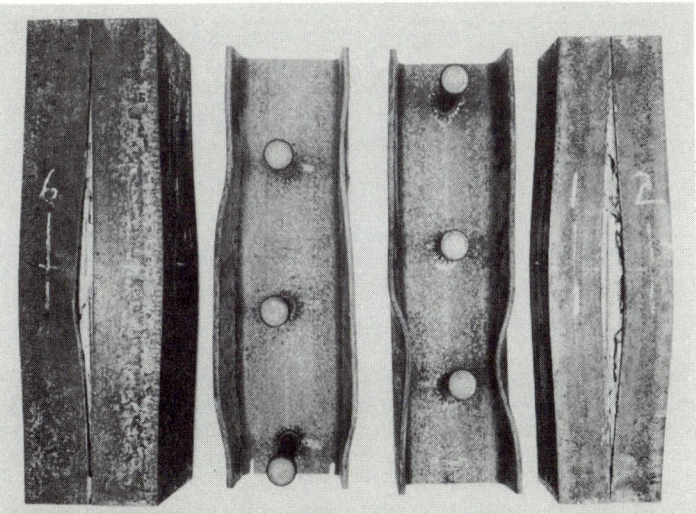

Fig. 2.2. Test specimens used to obtain the strength of the concrete contained by the channels; (failure of the specimens by the channels bowing outwards could not happen in a beam so the results may not be sufficiently representative of actual construction)

connectors. In this regard composite reinforced concrete is different to normal reinforced concrete. Although the part of the concrete in the compression zone of reinforced concrete which is contained by the shear links is effectively restrained, that concrete providing the cover to the reinforcement, and thus the most highly stressed part of the compression zone, is not. As a result the enhancement of the concrete within the links in normal reinforced concrete has no overall beneficial effect. The fact that in composite reinforced concrete the concrete near the soffit of the beam in a region of hogging moment will be effectively restrained is felt to be an important advantage over normal reinforced concrete.

Test results

The results of the University of Manchester tests are summarized in Table 2.1. The carrying capacity of the concrete contained within the channels of a specimen was determined as the difference between the strength of the specimen and that of similar pairs of channels without concrete. These carrying capacities have been translated into the

Table 2.2. Details of Asaad's 'hogging type' beams

Beam	Description of the longitudinal tension reinforcement*	Description of the shear reinforcement†
1A	under-reinforced	CP 110
1B		50% of CP 110
2A	balanced	CP 110
2B		50% of CP 110
3A	over-reinforced	CP 110
3AA‡		CP 110

*The term 'balanced' indicates an approximate balance between the strength of the tensile reinforcement and the combined compressive strength of the steel channel and the confined concrete; under or over-reinforced indicates less or more tensile reinforcement than the balanced amount.

†CP 110 indicates that the shear reinforcement was proportioned in accordance with the recommendations of CP 110.

‡Following the failure of beam 3A by longitudinal splitting of the flange, beam 3AA was made similar to 3A but with twice the amount of transverse reinforcement in the flange.

Beams	Tensile Reinforcement	Size of steel channel (mm x mm)
IA & IB	3 – 20 mm 4 – 8 mm } H.T. bars	152 x 89
2A & 2B	7 – 20 mm H.T. bars	152 x 89
3A & 3AA	7 – 20 mm 2 – 10 mm } H.T. bars 2 – 12 mm strand	152 x 76

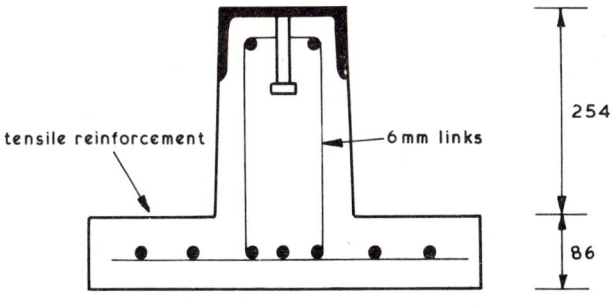

tensile reinforcement

6 mm links

254

86

Typical Section

Fig. 2.3. Details of Asaad's hogging type beams

strength of concrete as a proportion of the cube strength. As might be expected the results show considerable scatter, and to some extent the results depend on the size of the channels. However there are some generalizations that can be made.

(a) The combined compressive strength of concrete and channels is significantly higher than that of the channels acting alone, the difference being sufficiently high to be taken into account in design.

(b) The effective carrying capacity of the concrete is significantly influenced by the transverse reinforcement.

(c) The strength of the concrete in which links were incorporated ranged from $0.75 f_{cu}$ to $1.1 f_{cu}$ for the specimens with practical size steel channels.

(d) The mean longitudinal strains measured at or near the ultimate load were normally well in excess of the value of 3.5×10^{-3}, the strain usually associated with the rupture of concrete.

Discussion of results

The tests reported here are by no means exhaustive and more need to be carried out, particularly on specimens made with the larger size of channels. However an important question arises. Are these compression tests sufficiently representative of the conditions in beams? For example, failure of the compression specimens is as a result of the bowing out of the two channels in opposite directions (see Fig. 2.2). The situation in a beam is that the curvature of the channel due to a hogging bending moment is such as to make the channel tend to bow in towards the web at failure, a tendency which is restrained by the beam. There is an intuitive feeling that in consequence the concrete within the channel of a beam would have much greater restraining forces near the stage of collapse than in the compression specimens, resulting therefore in higher strengths and higher strain capacities. It would seem desirable to test further compression specimens in which the channels have an initial curvature which would eliminate the tendency to bow outwards.

Notwithstanding the scientific interest in the strength of the concrete contained within the channel, it is not likely to be of paramount importance in design. In practical situations the compressive force carrying capacity of the concrete will be smaller than that of the steel channel and associated compression reinforcement. However, the high strain capacities of the concrete, together with the resulting forces restraining the buckling of the channel, are likely to prove of considerable value in the ultimate load behaviour of continuous beams.

TESTS ON SOME 'HOGGING TYPE' BEAMS

Prior to testing continuous beams two series of tests were carried out on simply supported beams simulating the hogging moment region of continuous beams. The aim was to obtain data relating to

(a) the ultimate resistance moment of a hogging moment section

(b) the maximum compressive strains at failure
(c) the rotational capacities of the region adjacent to the maximum moment
(d) the influence of shear reinforcement on strength and rotational capacity
(e) moment–curvature relationships
(f) the possibility of utilizing compression reinforcement in conjunction with the steel channels.

Series 1

Six inverted T-beams (three pairs) were tested by Asaad[1] as simply supported beams. Details of the beams are summarized in Fig. 2.3 and Table 2.2.

One pair of beams had what, for the purpose of this discussion, will be described as 'balanced reinforcement' in that there was an approximate balance between the tensile strength of the tension reinforcement and the estimated combined compressive strength of the steel channel and the concrete contained within the channel. Of the other two pairs of beams, one was 'under-reinforced' and the other 'over-reinforced'. (In the light of the tests the beam with balanced reinforcement turned out to be slightly under-reinforced.)

It had been intended to study the influence of shear by proportioning the shear reinforcement in one beam of each pair in accordance with CP 110 and by giving the other beam 50% of this amount of shear reinforcement. In the event two pairs were in accordance with this, but the first of the over-reinforced beams to be tested failed by longitudinal splitting along the junction of flange and web. As a result the second over-reinforced beam became a repeat of the first but with more transverse flange reinforcement.

The arrangements for testing are shown in Fig. 2.4. The beams, of total length 3·5 m, were simply supported over a span of 3 m and subjected to a point load at midspan. The flange of the beam was at the bottom and the channel at the top. Thus the arrangement is an inverted version of the normal situation for a hogging moment region of a continuous beam. Longitudinal strains on both steel channel and concrete were measured in the midspan region using a Demec gauge of 100 mm gauge length. Several minutes were allowed to elapse after each increment to enable the beam to attain an equilibrium state before taking the strain readings.

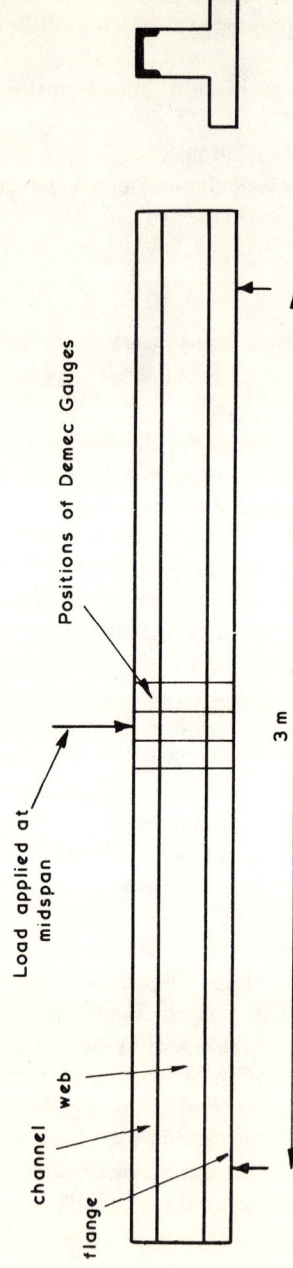

Fig. 2.4. Arrangement for testing Asaad's hogging type beams

The results of some of the tests are summarized in Figs 2.5 and 2.6 and in Tables 2.3 and 2.4 and the main points of interest are described below.

FAILURE MODE

The beams in which the shear reinforcement was in accordance with CP 110 (together with adequate flange reinforcement) failed in a typical flexural mode with wide flexural cracking at and near midspan as a result of yielding of the reinforcement. However, the beams were untypical of normal reinforced concrete in that there was no sign of the compression side of the beam failing. Indeed complete collapse of these beams did not occur, the tests being stopped when the deflexion began to increase at the same load. The compression zone always remained intact, with no sign of buckling of the channel (despite the high strains) and, when parts of the channel at midspan were subsequently removed, with no sign of crushing of the confined concrete.

The two beams in which the shear reinforcement was only 50% of the CP 110 requirement failed as a result of the shear cracking adjacent to midspan, although the load causing this failure was only moderately lower than the beams more heavily reinforced in shear.

ULTIMATE LOAD

Two methods of calculating the bending moment to cause failure of the beam have been adopted: an equilibrium method and a strain compatibility method.

The first is similar to the simplified method of CP 110 and simply equates the yield force in the tensile reinforcement to the compressive force in the compression zone in order to find the depth of the latter and hence the lever arm between the forces. The assumptions made in this method of calculation were that the yield stress of the reinforcement equals the 0·2% proof stress, that the channel yields in compression above the calculated neutral axis and any tensile stress in the channel below the neutral axis can be ignored, and that the compressive strength of the concrete contained within the channel equals $0·9 f_{cu}$. This value for the strength of the concrete is 50% higher than that for concrete in a normal laboratory beam and was decided on from a study of the tests on the compression specimens reported earlier and also of the beam tests. Calculations of the resistance moment of beam 2A using

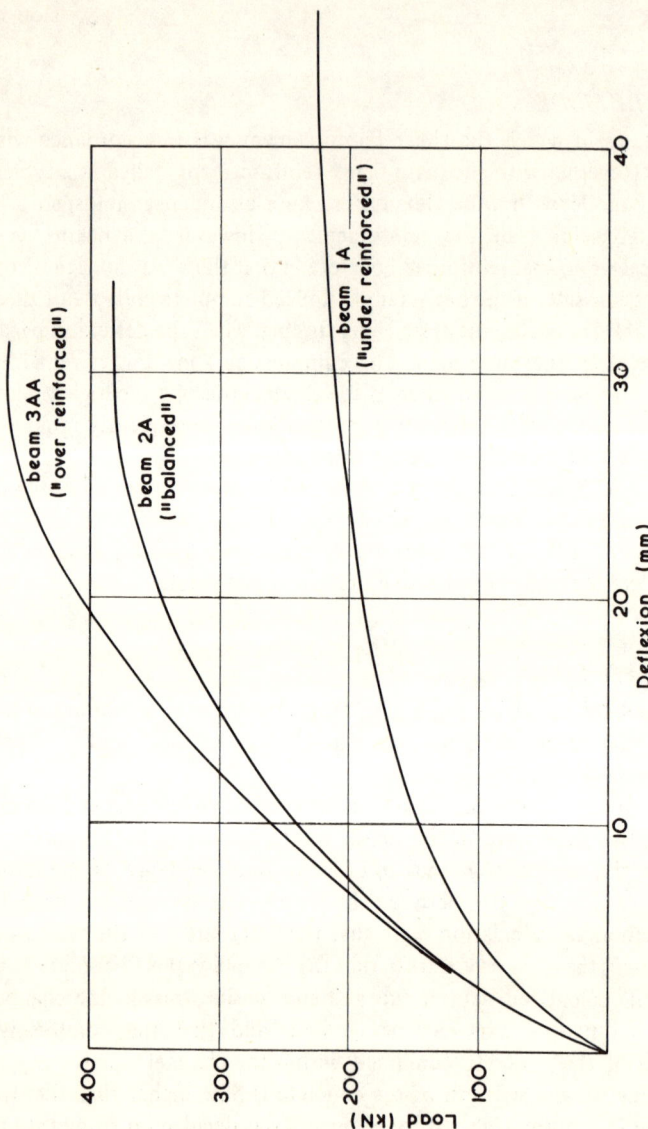

Fig. 2.5. Load–deflexion curves for some of Asaad's hogging type beams

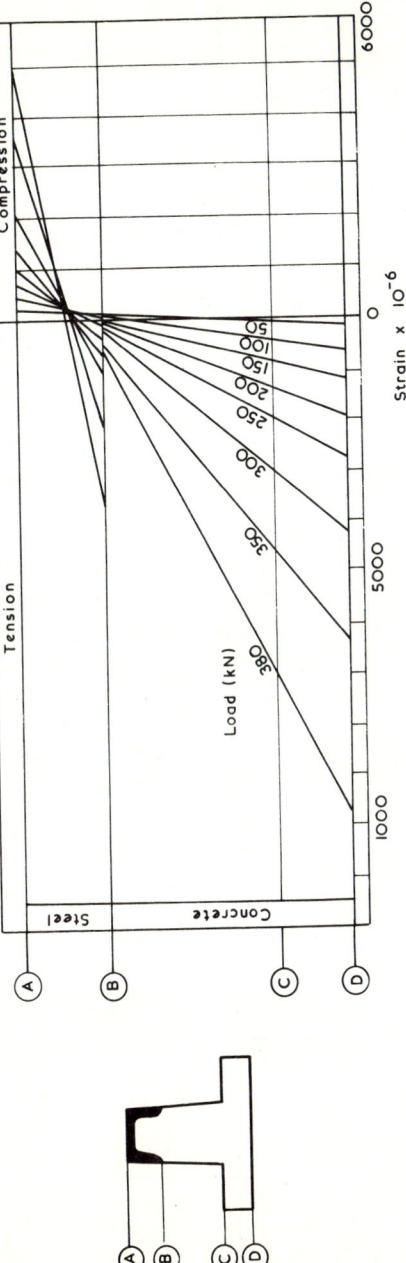

Fig. 2.6. Strain distributions at mid-span of Asaad's beam 2A

Table 2.3. Results of Asaad's hogging-type beams: beams failing in flexure

Beam	Theoretical ultimate resistance moment calculated by		Experimental bending moment at the maximum load, M_e: kNm	Ratio: M_e/M_{C1}	Ratio: M_e/M_{C2}	Depth of compression zone calculated by	
	the equilibrium method, M_{C1}: kNm	the compatibility method, M_{C2}: kNm				the equilibrium method (theoretical): mm	the strain diagram (experimental): mm
1A	143	160	173	1·2	1·08	25	40
2A	250	275	284	1·14	1·04	65	50
3AA	323	370	350	1·08	0·95	120	110

Table 2.4. Results of Asaad's hogging-type beams: beams failing other than in flexure

Beam	Type of failure	Experimental bending moment at failure: kNm
1B	Shear	158
2B	Shear	248
3A	Flange splitting	270

this method are given in Appendix 2 and the calculated values for the beams failing in flexure are summarized in Table 2.3. A comparison with the experimental values for the beams failing in flexure shows that the theoretical values are always lower, the ratios of M_e/M_{cl} ranging from 1·08 to 1·2.

The main reason for the discrepancies between the experimental bending moment causing flexural failure and the calculated values using the equilibrium method is that the strain hardening of the tensile reinforcement is not taken into account. The strain in the reinforcement at failure is somewhat higher than that corresponding to the 0·2% proof stress. Accordingly the stress would be slightly higher than that adopted in the calculation.

A method of calculation using strain compatibility considerations as in CP 110 would be expected to give more accurate results than the equilibrium method. However such a method depends on knowing the ultimate compressive strains and the strength of the contained concrete in compression. As a first step to providing some data on these, the resistance moments have been calculated using measured strains to obtain the stress in the reinforcement at the ultimate load and the approximate position of the centroid of compression. The resistance moments so obtained are summarized in Table 2.3 which shows that there is quite good agreement between the theoretical and experimental values.

STRAINS

The maximum compressive strains measured on the channels for the three beams failing in flexure were 0·004, 0·005 and 0·0085. In considering these values it must be remembered that the tests were stopped prior to any obvious distress of the concrete. In retrospect the tests were stopped too early. In any further tests the loading should be continued until there is a distinct falling off of the load-carrying capacity.

Series 2

Further tests on six simply supported inverted T-beams were made by Najmi.[2] The principal aim was to investigate the possibility of incorporating high yield reinforcing bars within the channel to act with the

Beams	Tensile Reinforcement	Compression Reinforcement
SO	6 − 20 mm	nil
S2	8 − 20 mm	2 − 20 mm
S4	10 − 20 mm	4 − 20 mm

All longitudinal reinforcement in H.T. bars

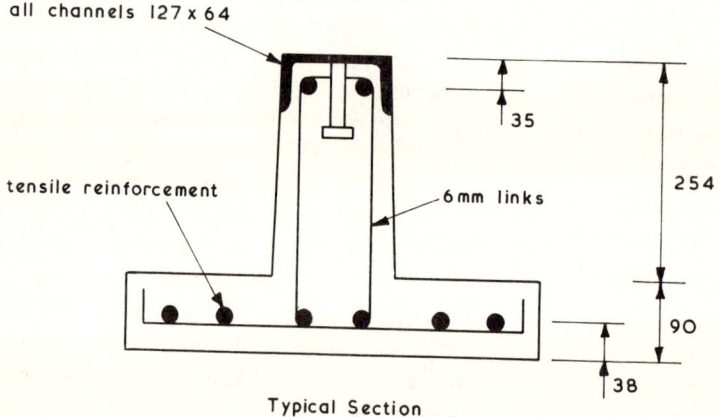

Fig. 2.7. Details of Najmi's hogging type beams

channel in resisting the compressive forces. An additional aim was to try to establish the criteria determining the stage of failure of the beams. In this respect it will be recalled that the series 1 tests were stopped too early and prior to any falling away from the maximum load.

Details of the beams are summarized in Fig. 2.7. All beams were reinforced to have an approximately 'balanced' section. A further difference compared with series 1 was that the span over which the beams were loaded was 4 m. A discussion of the results will be confined to the beams having 100% CP 110 shear reinforcement.

All beams showed the remarkable facility observed in series 1 of high rotational capacity at the ultimate load. The maximum compressive strains measured in these beams was of the order of 0·03. (In order to confirm this particular value it can be re-stated as nearly nine times the normally accepted mean maximum strain of concrete of 0·0035.) Again these incredibly high strains were attained without any sign of compression failure of the concrete contained by the channel, as was clear when parts of the channels were subsequently removed.

Soon after such high strains had been attained some slight buckling of the channels adjacent to midspan could be detected, although this was also accompanied by the lateral buckling of the whole beam, despite precautions to prevent this. At such stages the tests were suspended.

A typical load—deflexion diagram for a beam of this series is shown in Fig. 2.8. Two values of theoretical collapse load are also indicated on the diagram, both these values being lower than the actual collapse load. (The calculations for these theoretical collapse loads are given in Appendix 2.)

The higher of the two theoretical values corresponds to the collapse load calculated assuming a balanced section and based on the tensile forces. The reason for the discrepancy between this calculated value and the actual value is that the stress in the reinforcement at failure was assumed to be the 0·2% proof stress, whereas the extremely high strains attained at the level of the tension reinforcement would have meant appreciably higher stresses being reached.

The other theoretical collapse load is that based on the compression forces. There are several reasons which might explain why this theoretical value is lower than that based on the tensile forces.

(a) The beam was slightly over-reinforced in that the depth of the compression zone was slightly deeper than the channel. This extra compressive force has been ignored.

(b) The actual strength of the confined concrete may be somewhat higher than the assumed value of $0·9 f_{cu}$.

(c) The channel had reached such high compressive strains that the greater part of the steel channel would have reached the strain-hardening region of the stress—strain relationship.

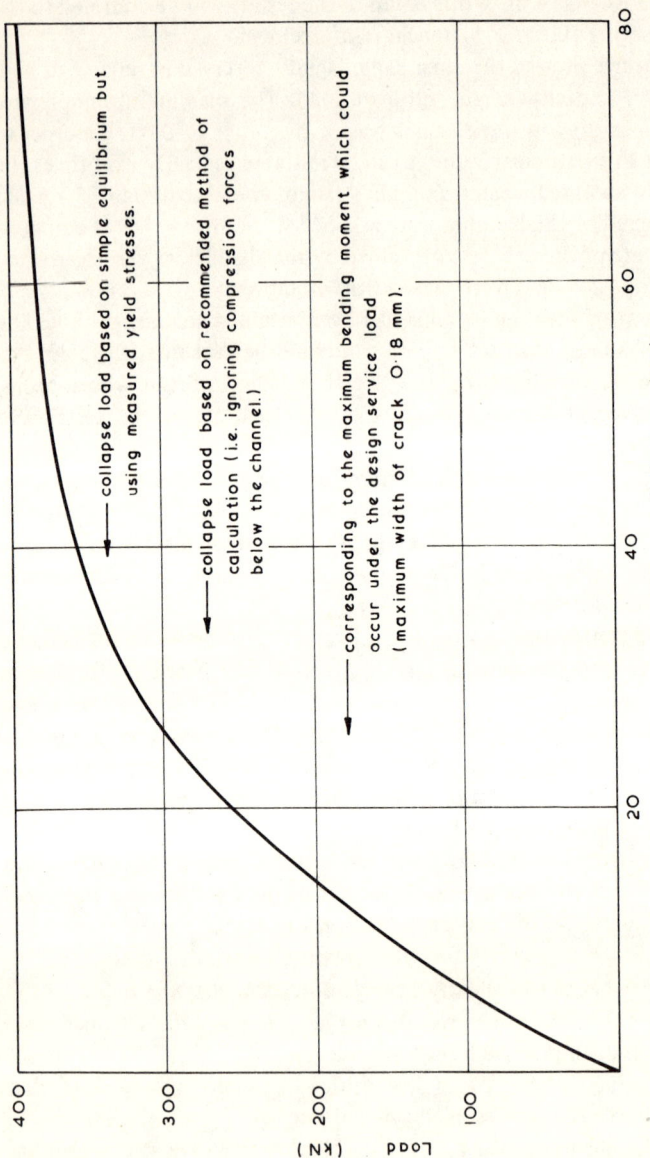

Fig. 2.8. Load–deflexion curve for Najmi's beam S2F

(d) The compression reinforcement had also reached strains corresponding to stresses greater than the 0·2% proof stress.

Despite the very high flexural strains attained in the tests in both tension and compression, it will not be practical to allow for strain hardening in practice. Simple methods of calculation must be adopted although this will undoubtedly lead to under estimating actual strength.

CALCULATION OF ULTIMATE HOGGING RESISTANCE MOMENT

The previous test results indicate that an abbreviated form of the previous equilibrium method is all that is required for the analysis of a given section for normal circumstances. It is suggested that the first step is to determine whether the beam is under-reinforced, balanced, or over-reinforced. A beam is considered to be balanced if

$$F_r = F_{ch} + F_{cc}$$

where the forces in the reinforcement, channel and the contained concrete are defined below. A beam is under-reinforced if

$$F_r < F_{ch} + F_{cc}$$

and over-reinforced if

$$F_r > F_{ch} + F_{cc}$$

For an under-reinforced beam (as previously defined) the strength is based on the tensile forces, thus

$$M_h = F_r z_{ch}$$

where F_r is the yield force of the tension reinforcement (which in the case of design incorporates the partial safety factor for the reinforcement), and z_{ch} is the lever arm assumed as the distance between the centroid of the tensile reinforcement and the centroid of the channel.

For a balanced or over-reinforced beam the strength is based on the compressive forces, thus

$$M_h = F_{ch} z_{ch} + F_{cc} z_{cc}$$

where F_{ch} is the yield force of the channel, F_{cc} is the maximum possible compression force in the concrete contained within the channel (at this stage assumed as $0·9 f_{cu}$ but in the case of design would include the partial safety factor to give $0·6 f_{cu}$) and z_{cc} is the distance between the centroid of the concrete contained by the channel and the centroid of the tensile reinforcement.

This calculation does not differentiate between balanced and over-reinforced beams. For the over-reinforced beams the calculation has ignored the effect of the concrete in compression outside the channel. This is similar to the calculation given in CP 110 for reinforced concrete when the compression zone is deeper than half the effective depth.

For practical beams it will sometimes be necessary to incorporate reinforcing bars within the steel channel in order to enhance the compressive strength. Such beams will, of course, be balanced or over-reinforced and the calculation of the ultimate resistance moment will therefore be based on the forces in compression. Thus

$$M_\mathrm{h} = F_\mathrm{ch}\, z_\mathrm{ch} + F_\mathrm{cc}\, z_\mathrm{cc} + F_\mathrm{rc}\, z_\mathrm{rc}$$

where F_rc is the yield force of the compression reinforcement and z_rc is the distance from the centroid of the reinforcement in compression to the centroid of the tension reinforcement.

DESIGN OF CONTINUOUS BEAMS

The relatively high hogging moments over the supports in continuous beams are a severe constraint in the design of concrete structures. These hogging moments are usually higher than the sagging moments in the span for the case of elastic bending moment diagrams or elastic bending moment envelopes. Such ratios of hogging to sagging are undesirable for the proportioning of sections in T-beam construction since, whether it be composite reinforced concrete, reinforced concrete or composite construction, the sagging resistance moment has a greater potential strength than the hogging resistance moment. It is therefore desirable from the economic standpoint to take advantage of any moment redistribution that can occur at the ultimate load to alter the ratios as much as possible. Composite reinforced concrete has the advantage over the traditional concrete materials regarding the rotational capacities of 'hinge' regions.

In the design of reinforced concrete continuous beams it is usual to start from the elastic bending moment envelopes corresponding to the ultimate load. This is then amended by allowing moment redistribution and the maximum hogging bending moment would usually be reduced. The maximum permissible degree of moment redistribution is 30%, although this is also dependent on the depth of the compression zone —

the deeper this zone the smaller the permitted degree of moment redistribution. This often places a severe restriction on the degree of moment redistribution from hogging moment regions of reinforced concrete T-beams since it is here that the compression zone tends to be comparatively deep, at least when one is attempting to use beams of minimum overall depth.

As already explained, composite reinforced concrete has some advantages over reinforced concrete in this latter respect. The comparatively large area of compression steel in the form of the channel together with the area of confined concrete within the channel enable high compression forces to be taken by a relatively shallow compression zone, so increasing the degree of rotational capacity of the critical sections in the hogging moment regions. Thus, in well designed beams of composite reinforced concrete, it is always possible to develop a high degree of plasticity in hogging moment regions at the ultimate load.

However, the maximum amount of moment redistribution in composite reinforced concrete is determined by the serviceability of the beam. In this respect it must be remembered that the tensile forces in regions of hogging bending moments are taken by normal reinforcing bars, usually high yield bars. Since in normal reinforced concrete the minimum ratio of the ultimate resistance moment of the section and the service load bending moment is approximately 1·25 (this being the average global safety factor, 1·75, reduced by the 30%), this ratio could presumably be adopted for the hogging moment sections of composite reinforced concrete. This should ensure that the tensile cracking in the concrete slab over the supports is not excessive at the service load.

For the design of beams of composite reinforced concrete it is therefore suggested that the elastic bending moment diagram corresponding to the service load is the starting point for design. The ultimate resistance moment of the hogging moment section at the support should then be made equal to at least 1·25 times the service load bending moment. The bending moment diagram for the ultimate load case can then be drawn by assuming that plastic analysis will apply, so determining the required value of the resistance moments of the sagging moment region.

Despite this procedure, in many practical cases the ratio of the ultimate hogging resistance moment to the ultimate sagging resistance moment remains undesirably high. As a result one finds that, if one determines the size of steel channel from the required hogging resistance moment, the amount of strand reinforcement then required to act with the channel to resist the sagging bending moment is low. Whilst from the technical standpoint this is quite acceptable, it is not desirable from the economic standpoint since the greater the proportion of strand the more economical the section. To overcome this it is proposed that, in proportioning the hogging moment section, use be made of compression reinforcement acting within the steel channel. This will enable a smaller channel to be adopted which, in the sagging moment region, would then require a higher proportion of strand to resist the tensile forces.

One further point to be remembered when considering the behaviour of continuous beams of composite reinforced concrete under loads increased gradually to failure is that non-linear behaviour begins soon after the service load has been reached. Yielding of the channel starts in the region of the sagging bending moments at, or soon after, the service load and spreads along the beam. For a beam subjected to a uniformly distributed load the full region of yield at the ultimate load

Fig. 2.9. A continuous beam under test (the discontinuities in the line of the beam due to the plastic regions at the positions of the first load and the central support can be observed)

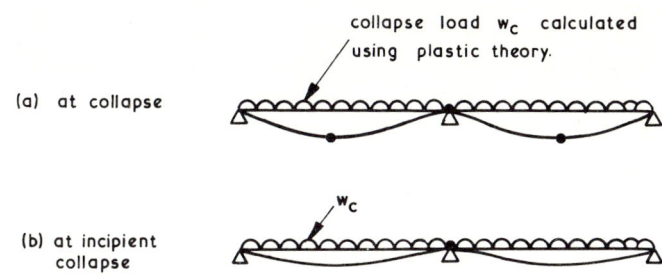

Fig. 2.10. Stages in the plastic collapse of a continuous beam

can be quite extensive (see next chapter, Fig. 3.3). The main implication of this for continuous beams relates to the demands made on the rotations of the plastic hinges forming in hogging moment regions.

Plastic hinge requirements

The results of the tests on the simply supported beams for both sagging and hogging type curvatures have indicated that hinge regions for both types of curvatures would have very high rotational capacities. This immediately suggests that beams of composite reinforced concrete can be designed assuming that plastic theory can be applied without restriction. Such a proposal needs thoroughly investigating, both experimentally and analytically, and further work in this field is desirable.

Some continuous beams in composite reinforced concrete have in fact been tested at the University of Manchester (Fig. 2.9) and all reached their full predicted collapse load and collapsed by the formation of hogging moment and sagging moment plastic hinges. The incredible plastic behaviour of the hogging moment regions was again demonstrated, and on the evidence of these tests there is little doubt that plastic design could be adopted. However to confirm this finally, it will be necessary to analyse a wide ranging set of practical designs. In order to facilitate such analysis, aspects of the problem are now discussed.

The method of calculating the required rotation of plastic hinges in reinforced concrete has been developed by A. L. L. Baker.[3] The method can be illustrated by reference to the diagrams in Fig. 2.10. Fig. 2.10(a) shows the stage of collapse of a two span continuous beam by the formation of plastic hinges. These hinges do not form at the same time.

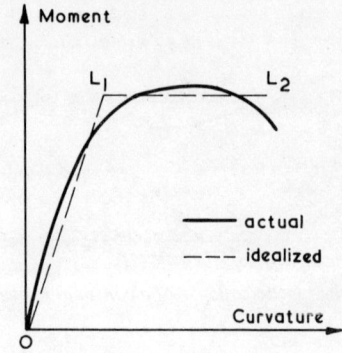

(a)   A   Reinforced   Concrete   Section

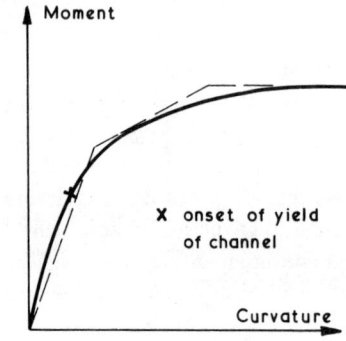

(b)   A   Composite   Reinforced   Concrete

Section   in   a   sagging   moment   region.

Fig. 2.11. Typical moment–curvature relationships

For example, if the hinge at the central support forms first, the situation at incipient collapse would be as shown in Fig. 2.10(b) where the sagging moment hinges are just about to form. At this stage there is an angular discontinuity between the beams meeting at the central support due to the rotation of the plastic hinges. For a given beam the plastic resistance moment at the support is known and the situation illustrated in Fig. 2.10(b) is statically determinate. As a result the amount of hinge rotation from the initial formation of the hinge to the collapse state can be easily calculated. This is known as the required rotation.

For normal reinforced concrete this calculation is straightforward as it can be assumed that the EI value is constant for the regions outside the plastic hinges. A typical moment–curvature relationship for a reinforced concrete section is illustrated in Fig. 2.11(a); it is only slightly curving until yield of the reinforcement occurs. At this stage the curvature increases much more rapidly and indeed the ultimate resistance moment of the section is reached soon after the onset of yield. For practical purposes a simplified bi-linear curvature diagram can be assumed with little error. This is shown in Fig. 2.11(a) where the straight line OL_1 represents the idealized elastic phase for the regions outside the initial hinges.

Unfortunately this simplification cannot be applied to composite reinforced concrete. It will be recalled that in regions of sagging bending moments yielding of the channel begins soon after the service load stage and so much earlier in relation to the ultimate load than for reinforced concrete. A typical moment–curvature relationship for composite reinforced concrete is shown in Fig. 2.11(b) where it will be clear that a bi-linear simplification would not be applicable.

The effects of this early yielding of the channel in the sagging moment region are twofold, the first being that the required rotations at hogging moment plastic hinges of composite reinforced concrete are increased. The second effect is that the calculation of the required rotation of the plastic rotations of plastic hinges is rather more complicated than for reinforced concrete. For example, whereas in reinforced

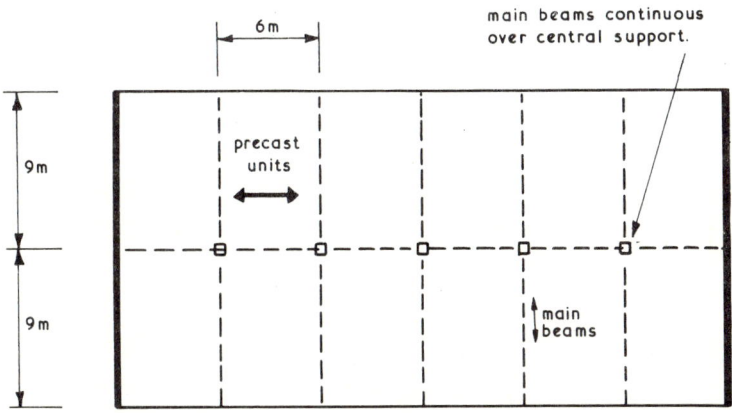

Fig. 2.12. Floor plan used for illustrative designs of continuous beams

concrete the calculation normally requires the product integration of standard bending moment diagrams having parabolic and straight line forms and therefore is straightforward, the calculation for composite reinforced concrete requires the product integration of non-standard diagrams.

Nevertheless, despite the greater tedium of the calculation, it does remain possible, although for wide ranging calculations it will be necessary to develop a suitable computer program. In this regard work at the University of Manchester has not yet been completed.

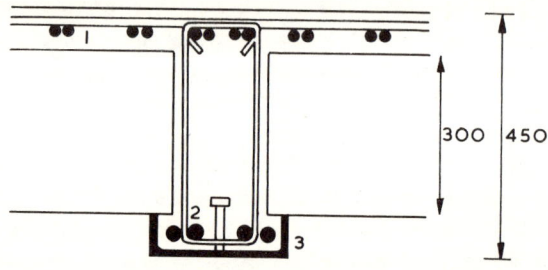

1.   20mm high yield bars

2.   32mm high yield bars

3.   Channel 254 x 76mm

(a)   hogging moment section

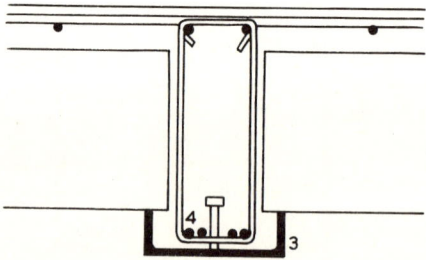

4.   2 − 12mm and 2 −15mm strands.

(b)   sagging moment sections

Fig. 2.13. Cross-sections of the beam in the continuous beam design example

Design examples

For the purpose of illustrating some of the previous points, examples of the design of composite reinforced concrete involving continuity are now presented. It has been assumed that plastic theory can be applied.

TWO SPAN CONTINUOUS BEAM

The arrangements and loadings are similar to those in the design example in chapter 1 except that the building is considered to be 18 m wide and having two spans of 9 m with continuity of the channels over the central supports (Fig. 2.12). As a result of the continuity it will be possible to reduce the depth of construction from that used for the simply supported spans of the designs B in chapter 1. An overall depth of 450 mm has been adopted giving a span to depth ratio of 20. Cross-sections of the beams are shown in Fig. 2.13. The combination of hollow precast units in lightweight concrete and the in situ concrete topping are equivalent from the dead load standpoint to a solid slab of 200 mm in depth.

Hogging bending moment

Characteristic dead load	= 38 kN/m
Characteristic imposed load	= 24 kN/m
Design load for serviceability limit state	= 62 kN/m
Design load for ultimate limit state	= 91 kN/m
Maximum hogging moment at service load (S.L.)	
$= 0.125 \times 62 \times 9^2$	= 628 kNm
Minimum resistance moment at support $= 1.25 \times 628$	= 785 kNm

Hogging resistance moment

Reference to Fig. 2.13 shows that the beam has been designed incorporating reinforcing bars within the steel channel to act in compression.

(a) Channel area   = 3603 mm²
    Design yield force in channel   = 840 kN
    Lever arm = 450 − 41 − 18   = 391 mm
    Resistance moment provided by channel   = 329 kNm

(b) Area of compression reinforcement   = 3217 mm²
    Design yield force = 3217×0.312 = 1000 kN
    Lever arm = 450 − 41 − 39   = 370 mm

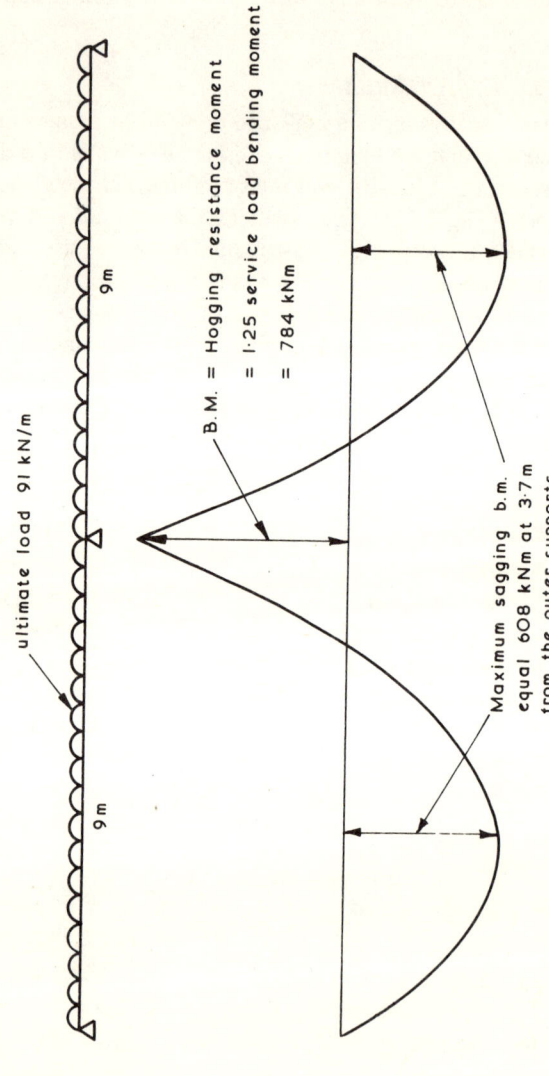

Fig. 2.14. Bending moment diagram corresponding to the ultimate load of the beam in the continuous beam example

	Resistance moment provided by compression bars	= 370 kNm
(c)	Area of concrete contained in channel	= 12500 mm^2
	Design force = 12·5 × 0·6 × 30	= 225 kN
	Lever arm = 450 − 41 − 42	= 368 mm
	Resistance moment provided by concrete	= 85 kNm
	Total hogging resistance moment	= 784 kNm

The reinforcement required on the tension side of the beam to resist the total compression force is 18—20 mm dia. high yield bars.

Sagging bending moment

The bending moment diagram for the ultimate load is shown in Fig. 2.14. It is determined by assuming a moment over the supports equal to the hogging resistance moment. This determines the maximum sagging moment at the ultimate load as 608 kNm which occurs at 3·7 m from the outer supports.

Sagging resistance moment

Design yield force in channel	= 840 kN
Design ultimate force in strand	= 682 kN
Total design tensile force	= 1522 kN
Effective width of flange	= 2·05 m
Depth of compression zone	= 62 mm
Lever arm for channel	= 401 mm
Resistance moment from channel	= 338 kNm
Lever arm for strand	= 388 mm
Resistance moment from strand	= 266 kNm
Total sagging resistance moment	= 604 kNm

Service load stresses in sagging moment region

To check service load stresses it is necessary to obtain the bending moment envelope corresponding to the service load. This indicates that the maximum sagging bending moment is 402 kNm at 3·6 m from the support.

Approximate maximum steel force = 402/0·399   = 1006 kN
Total steel area   = 4068 mm²
Average stress in steel at the service load   = 247 N/mm²
Maximum stress in channel   = 251 N/mm²

FIXED ENDED BEAM

The arrangements and loads are assumed to be identical to the previous case except that both ends of the 9 m span beam are assumed to be fixed. The procedure is similar to the previous case and results in the

1.   18 - 20 mm high yield bars

2.   4 - 25 mm high yield bars

3.   178 x 76 channel.

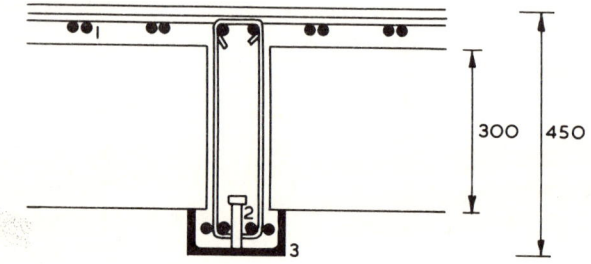

(a) hogging moment section

4.   2 — 15 mm strands

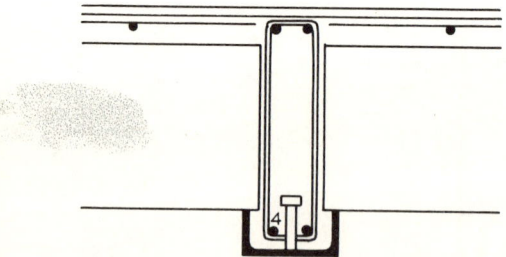

(b) sagging moment section

Fig. 2.15. Cross-sections of the beam in the fixed ended beam design example

cross-sections shown in Fig. 2.15. A summary of the calculations is given.

Maximum hogging moment at S.L. = $(62 \times 9^2)/12$	= 420 kNm
Minimum hogging resistance moment = $1 \cdot 25 \times 420$	= 524 kNm

Hogging resistance moment

Resistance moment from channel = $616 \times 0 \cdot 387$	= 238 kNm
Resistance moment from bars = $614 \times 0 \cdot 362$	= 222 kNm
Resistance moment from concrete = $160 \times 0 \cdot 368$	= 60 kNm

1.   12 − 20 mm high yield bars.

2.   2 − 32 mm high yield bars.

3.   203 x 76 channel

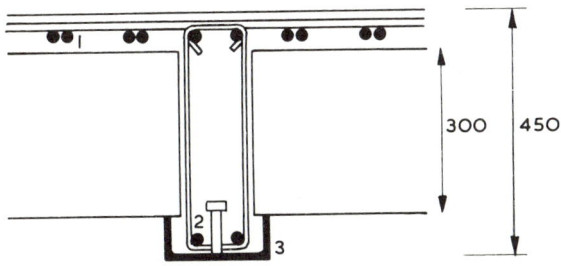

(a)   hogging moment section.

4.   2 − 12 mm strands (or 4− 16 mm h.y. bars)

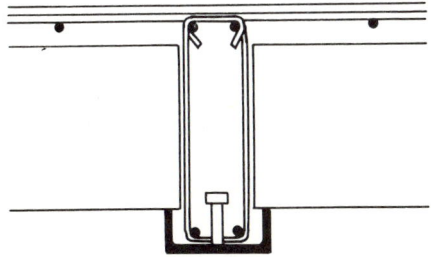

(b)   sagging moment section.

Fig. 2.16. Alternative cross-sections of beam in fixed-ended beam design example

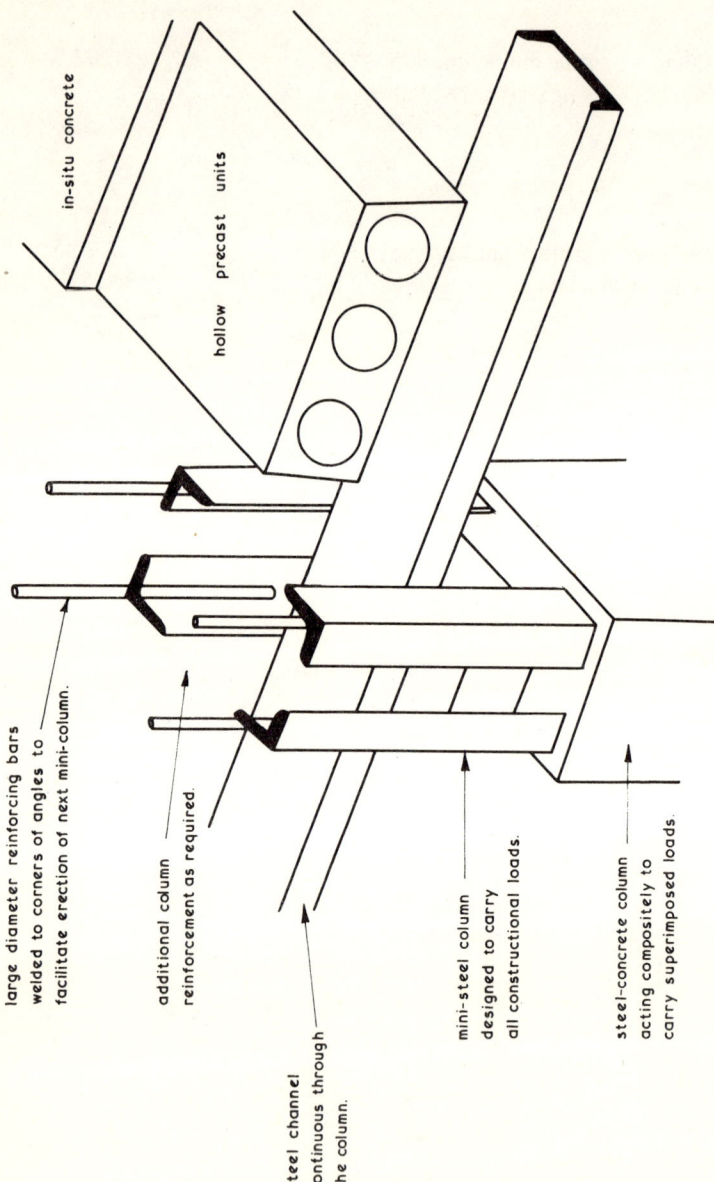

in-situ concrete

hollow precast units

large diameter reinforcing bars
welded to corners of angles to
facilitate erection of next mini-column.

additional column
reinforcement as required.

steel channel
continuous through
the column.

mini-steel column
designed to carry
all constructional loads.

steel-concrete column
acting compositely to
carry superimposed loads.

Fig. 2.17. Possible method of construction to obtain continuity of beams

Total hogging resistance moment   = 520 kNm

Sagging resistance moment

Sagging resistance moment required = $[(91 \times 9^2)/8]$ —
   520   = 401 kNm
Resistance moment from channel = 616×0.407 = 250 kNm
Resistance moment from strand = 396×0.40 = 158 kNm
Total sagging resistance moment   = 408 kNm

As a result of the changed end conditions compared with the two span continuous beam design the required resistance moments are reduced. As the same overall depth of beam was retained a smaller steel channel (178 mm $\times$ 76 mm) was required. Whilst this channel is acceptable for the size of beam, it would probably be regarded as the minimum suitable size from the standpoint of practical considerations such as providing sufficient seating for the precast units but retaining a sufficient width of in situ concrete between the precast units. Even for this size of channel it has been necessary to place two of the compression reinforcing bars outside the shear links. Although this would not be acceptable in normal reinforced concrete, it is acceptable in composite reinforced concrete as the channel restrains the compression bars from buckling. Indeed this position for the compression bars may well be the most convenient for steel fixing and could become the standard arrangement.

In the event that the size of channel chosen for the beam is dictated by the constructional problem, it may well become preferable to use ordinary reinforcing bars instead of the resulting small amount of prestressing strand. For example suppose a 203 mm $\times$ 76 mm channel had been chosen as the smallest possible channel for the particular circumstances. This would then have required only two 12 mm strands in the region of the sagging bending moment (Fig. 2.16). Whilst this would be quite acceptable from the technical standpoint, the designer might well decide to use the equivalent amount of ordinary high tensile reinforcing bars, in this case four 20 mm bars.

CONSTRUCTIONAL ASPECTS

Effecting continuity in steelwork and composite construction requires very costly connections. As a result continuity in these forms of construction is often avoided in favour of structures designed with simply supported beams. It is felt to be one of the advantages of composite reinforced concrete that continuity could be effected quite easily as long as the non-traditional approach is extended to the design of the columns and to the sequence of construction. There is much scope for constructional enterprise in this field and the following is just one such possible arrangement.

The mini-steel column is one in which steel sections of small size are used to carry the loads during the initial part of construction, but which are subsequently surrounded by concrete, the whole then acting as a reinforced concrete column. Extra steel in the form of reinforcing bars is incorporated in the concrete as required.

Although it has been suggested that the prior erection of a complete steelwork frame facilitates the subsequent construction of the floors, there are some disadvantages when precast flooring is used. The lifting of the precast units into place on a particular floor by the crane is considerably hampered by the steel frame above this floor. To overcome this it is suggested that the mini-steel columns are erected one floor at a time. This would also allow the beam channels to be lifted directly into place on to the cross-battens of the column and to be continuous through the column. This appears to offer a much cheaper way of effecting continuity than by the type of connection currently used in steelwork and composite construction.

An arrangement showing a possible type of mini-steel column is illustrated in Fig. 2.17.

SUMMARY

A composite reinforced concrete beam is well suited to resisting hogging bending moments, enabling very compact sections to be adopted. At the ultimate resistance moment of such sections extremely high rotational capacities can be attained.

Investigators at the University of Manchester believe that for all practical situations statically indeterminate structures could be designed in composite reinforced concrete assuming that plastic theory can be applied without restriction. However it is important to note that this cannot be said to have been verified absolutely and further proof, both experimentally and analytically, is required. To facilitate such further work, the type of studies still required are

(a) the development of a computer program for the generation of theoretical moment–curvature relationships for both sagging and hogging moment sections, to be verified by a comparison with test data

(b) the establishment of a method for calculating the theoretical rotational capacities of plastic hinges in both hogging and sagging moment regions, to be verified by a comparison with test data

(c) the development of a computer program for the analysis of continuous beams of composite reinforced concrete to determine the required rotations of plastic hinges

(d) the comparison of required rotations of plastic hinges with their rotational capacities for a wide ranging set of statically indeterminate structures.

Assuming the above belief will be verified, the suggested procedure for the design of a continuous beam is as follows.

(a) Calculate the hogging bending moments at the internal supports due to the full service load.

(b) Adopt the design ultimate hogging resistance moments as approximately 1·25 times the service load hogging bending moments and proportion the beam accordingly. (It will often be advantageous to incorporate reinforcing bars within the channel to help resist the compressive forces. This enables a smaller channel to be used which in turn enables more prestressing strand to be used in the region of sagging bending moments.)

(c) Draw the bending moment diagram corresponding to the ultimate load assuming that the moments at the internal supports equal the design ultimate hogging resistance moments.

(d) Determine the maximum sagging bending moments and calculate the amount of prestressing strand required to act with the previously proportioned steel channel to provide the design ultimate sagging resistance moments.

(e) Proportion the shear reinforcement in accordance with the recommendations given in chapter 3.

REFERENCES

1. ASAAD S. M. 'An exploratory investigation on the continuity problem for the new structural material composite reinforced concrete.' MSc Thesis, University of Manchester, 1977.

2. NAJMI A. 'A study of the ultimate load behaviour of continuous beams of composite reinforced concrete.' PhD Thesis, University of Manchester, 1979.
3. BAKER A. L. L. 'Limit state design of reinforced concrete.' Cement and Concrete Association, London, 1970.

CHAPTER 3

Problems of shear

EFFECTS OF VERTICAL SHEAR

It has already been shown that composite reinforced concrete is essentially a form of reinforced concrete as far as flexural behaviour is concerned. This is also true of its behaviour in shear. Nevertheless there are some differences which need to be highlighted. To appreciate these differences an understanding of the shear failure of normal reinforced concrete beams is required.

Shear behaviour of normal reinforced concrete

Under the action of a gradually increasing load a flexural crack develops in an inclined direction due to the effect of shear. At some stage in the loading, inclined cracks will extend from the level of the longitudinal reinforcement to the region of the compression zone. As such cracks attempt to develop and widen under the increasing load, the longitudinal bars produce a dowelling action across the crack. In beams without shear reinforcement this dowelling action tends to split the concrete horizontally along the level of the reinforcing bars, ruining the bond between the bars and the concrete and causing sudden failure.

However in beams with shear reinforcement, the sudden widening of the inclined crack is prevented. As the inclined crack extends and widens under increasing load, so more of the shear force is transferred from the concrete to the links crossing the inclined crack. Moreover the development of the horizontal splitting of the concrete along the longitudinal bars due to the dowelling action is now prevented by the links which envelop the bars, although horizontal cracking will develop to some extent.

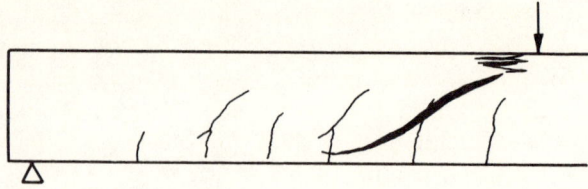

Fig. 3.1. Shear–compression type failure in a beam of normal reinforced concrete

Gradually, however, as the load on the beam increases, the inclined crack widens and extends up the beam penetrating into what was the flexural compression zone. As the area of the intact concrete above the inclined crack decreases, the stresses increase, and a stage is eventually reached when rupture of the zone occurs causing complete collapse of the beam. This rupture of the concrete has similarities to the rupture of the compression zone failing in flexure, and the mode of failure is therefore termed shear–compression (Fig. 3.1).

The dowelling action of the longitudinal reinforcement plays a significant role in the shear behaviour of the beam. In beams with shear reinforcement dowelling action carries part of the shear force across the inclined crack to the links beyond the end of the crack. Thus the links which are effective in carrying the shear force are more than those crossing a crack inclined at 45°.

Shear behaviour of composite reinforced concrete

Composite reinforced concrete has been developed for use as the main beams in framed structures. Such beams will always have shear reinforcement which can be expected to behave in a basically similar manner to that in normal reinforced concrete. Nevertheless there are several very important differences between the two types of beams which might be expected to have some influence on behaviour. A description of these differences, as far as regions of sagging moments are concerned, follows.

The steel channel, which forms a significant proportion of the longitudinal steel, is not enclosed by the shear links. It will be recalled that the enclosure of the longitudinal reinforcing bars by the shear links is an important facet of normal reinforced concrete. Indeed it is a requirement of CP 110. Clearly, however, the stud shear connectors on the channels will tend to produce the same effect. As explained in chapter

1, any tendency for the channel to separate vertically from the concrete web is restrained by the internal strut forces set up between the underside of the head of the stud and the longitudinal strand reinforcement (see Fig. 3.2). This interaction of channel, shear connectors, strand reinforcement and shear links is clearly of a complex nature, with perhaps the spacing of the stud shear connectors being a critical factor of the behaviour. Certainly the correct juxtaposition of stud head and strand reinforcement is of paramount importance.

In developing the dowel action of the reinforcement in normal reinforced concrete the lateral stiffness of the longitudinal bars is clearly significant. The greater this lateral stiffness the greater the ability of the reinforcement to carry shear force to more stirrups beyond the inclined part of the crack. Such behaviour is not explicitly allowed for in design but is certainly implicit in the code of practice recommendations since these are based on test data.

The channel of composite reinforced concrete clearly has considerable lateral stiffness, at least in comparison with reinforcing bars, and so has an apparent excellent ability to transmit dowel forces. The term 'apparent' is used since it must be remembered that the channel starts to yield at loads only slightly higher than the service load. As the load increases towards the ultimate load so the yield of the channel spreads through the depth of the channel and also along the channel. For a beam subjected to a uniformly distributed load the region of yielding of the channel is quite extensive, as illustrated in Fig. 3.3. This shows the region of full yielding at the ultimate load of the beam shown in Fig. 1.12 for which the flexural calculations were given in chapter 1.

It would seem logical that a channel fully yielding in tension would have little or no lateral stiffness and so little ability to transmit dowel

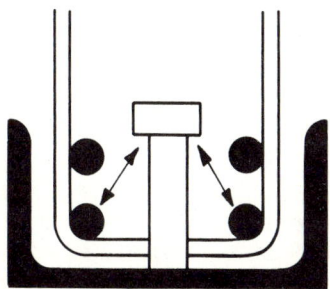

Fig. 3.2. Method of transfer of vertical forces from the channel to the links

forces. Fortunately, even if this is the case, the shear forces in this region of the beam are small so that it would be no economic hardship to have to design the shear links to be able to carry the whole of the shear force across a 45° crack without the benefit of dowel action.

In the outer regions of the beam illustrated in Fig. 3.3 where the shear forces are higher, the channel or part of it remains elastic and should be able to provide a dowelling action and so contribute to the shear strength of the beam.

The prestressing strand requires very high strains at the ultimate load. Although such strains have been shown to be attainable as far as flexure is concerned, the additional problem with shear is that at or near the ultimate load the cracking due to the dowelling action causes

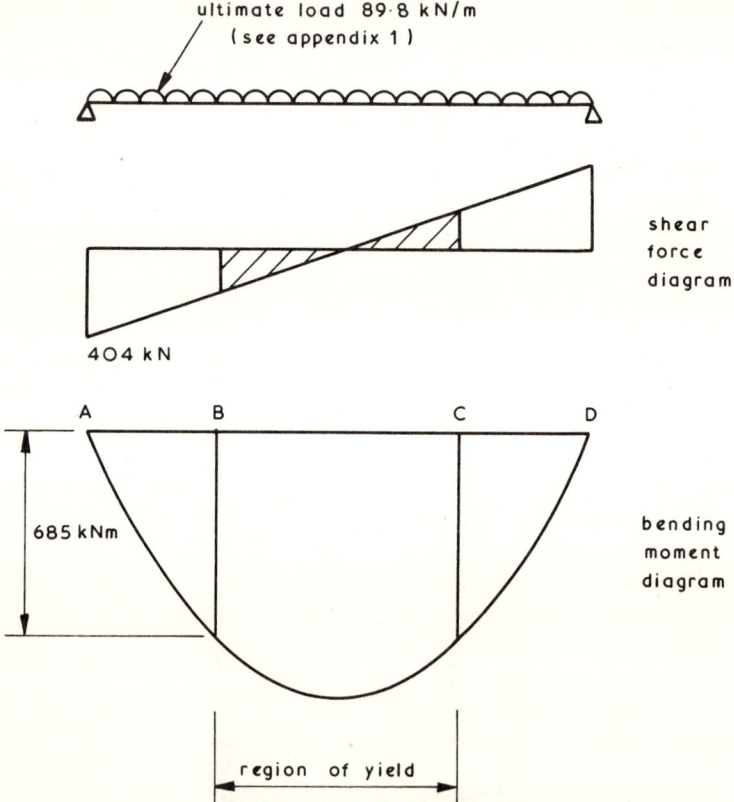

Fig. 3.3. Region of yield of channel at the ultimate load of beam in Fig. 1.12

loss of bond. The same effect occurs with normal reinforced concrete, but here the maximum strains required in the steel are much lower than those in composite reinforced concrete.

Again though, it is only in the central region of sagging bending moments that the very high strains are required in the strand. If, as suggested, these regions are designed for shear ignoring any contribution by dowelling action, there should be little cracking along the level of the reinforcement and hence little loss of bond. In the outer parts of sagging moment regions, where the shear forces are higher and dowelling action perhaps causes greater loss of bond, the strains required in the prestressing strand at the ultimate load are much lower.

Test results

The results of shear tests by Mills on 19 simply supported beams have been previously reported.[1] Unfortunately these tests were initiated whilst the reinforcing strand still being adopted in the tests was Bristrand, and only 3 of 19 beams were reinforced with prestressing strand. For full details of the tests reference should be made to the publication, but for convenience some of the important points are summarized.

The beams, of 5 m in length, had zones which were appreciably weaker in shear than the remaining parts of the beam. The idea was to obtain shear failure at positions

(a) away from the applied load so that enhancement due to arching action over the main shear crack could not occur

(b) which varied within the shear span to give different ratios of moment and shear.

The positions of the zones, termed critical zones, are illustrated in Fig. 3.4. It was expected that in zones A the channels would be yielding in flexure throughout their depth and so unable to contribute to any shear strength by dowelling action, whereas in zones B the channels were expected to contribute by dowelling action.

The tests surprisingly did not indicate any significant difference between the two shear zones. However it was found that, even in the zones A, the strains in the upper part of the flanges of the channel were often such as to be still elastic, and so presumably the channels were still contributing to shear strength. It seems possible that the few shear links in the critical zone allowed some loss of interaction between

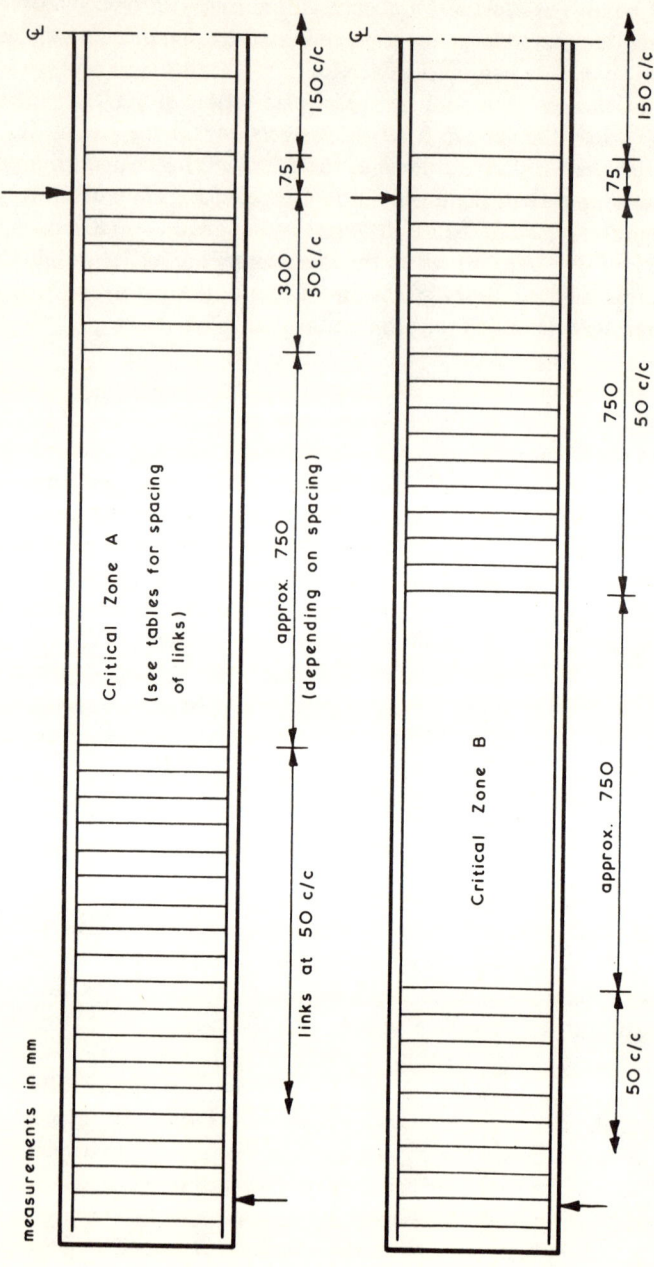

Fig. 3.4. Positions of the critical zones in the beams used in the shear tests (Mills)[1]

channel and the reinforced concrete resulting in an area of the channel remaining elastic. More tests of this kind are required to clarify the issue.

The beams were appreciably stronger in shear than had been expected. Many of the beams tested failed in flexure although the amounts of shear reinforcement were often appreciably less than would be required by CP 110. Indeed it was found that a spacing of links of at least twice that required in designs to CP 110 was required before a shear failure occurred, whilst some beams having a spacing of links 2½ times the CP 110 required spacing still failed in flexure. However, all the beams which failed in flexure were by then at incipient shear failure as evident from the wide inclined cracking.

The test results cannot be compared directly with reinforced concrete. It is possible that comparable reinforced concrete T-beams would have the same 'enhanced' strength in comparison with the requirements for shear in CP 110.

Whilst from the purely strength standpoint there might be a case for arguing for less shear reinforcement in beams of composite reinforced concrete than would be required by the recommendations of CP 110, the data obtained on the widths of the shear cracks at the service load would not support this. The width of the cracks in those critical zones

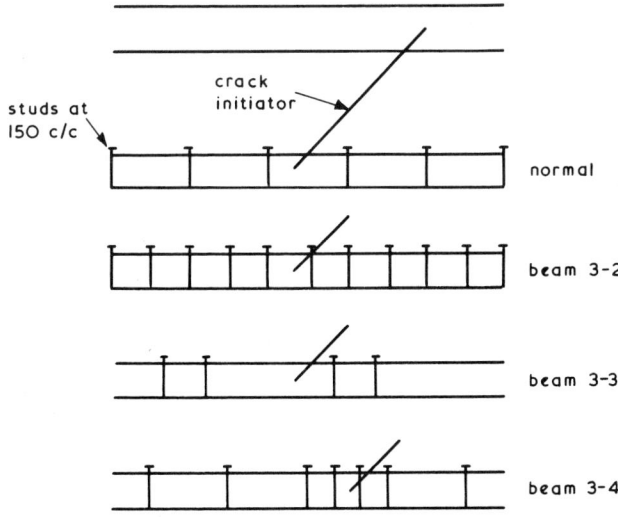

Fig. 3.5. Position of studs and crack initiators within the critical zones

with no special features (such as inclined crack initiators) generally attained values in the range 0·1 to 0·2 mm. It is to be expected that under sustained loading conditions these cracks would have increased in size.

Although cracks will not normally be visible in the envisaged method of construction, it is nevertheless felt that, from the serviceability standpoint, a spacing of links equal to that required by CP 110 may well be necessary. This is confirmed by the flexural tests described in chapter 1 which, while not reinforced exactly to CP 110, had comparable amounts of shear reinforcement.

The importance of the shear connector in preventing separation of the channel and the concrete and in providing the means for developing dowel action has been described earlier. To throw some light on this four beams were tested in which the only variable was the position and spacing of the studs within the critical zones. The details are given in Fig. 3.5. Within the range of spacing of the studs illustrated, there was no apparent effect. Although two beams failed in flexure and two in shear, all failed at approximately the same load.

PROBLEM OF HORIZONTAL SHEAR

Shear connectors

As in normal composite construction, shear connectors are required in composite reinforced concrete to transmit the horizontal shear between the steel section and the concrete. However it will be recognized that the conditions in the two types of construction have several important differences which will affect the behaviour of the shear connectors, the main differences being as follows.

(a) In regions of sagging bending, the shear connectors in normal composite construction will usually be embedded in concrete which is in compression, whereas in composite reinforced concrete the shear connectors will be embedded in concrete which is in a region of tension and therefore cracked.

(b) The interface of concrete and steel in normal composite construction will be close to the neutral axis and hence at a position of low steel stress, whereas the interface in composite reinforced concrete is at a position of maximum stress in the steel section.

(c) The stage of yielding throughout the whole steel section in normal composite construction marks the limit of the ultimate resistance moment of the beam, but that is not the case in composite reinforced concrete where complete yielding at a section of the steel channel can occur long before the ultimate resistance moment is reached.

Much research has been carried out on the use of shear connectors in normal composite construction. The strength of shear connectors is determined from push-out tests, details for which are laid down in the code of practice. The values so obtained from these tests are termed the 'characteristic strengths' of shear connectors. As in the case of concrete where the characteristic strength obtained from the cube test does not determine the strength of the concrete in a structure, because of the differing conditions applying in the two situations, the characteristic strength of shear connectors are not the precise strength of shear connectors in a beam. Factors, partly experimental but largely intuitive, are applied to translate the characteristic strengths into the design strengths. The factors recommended for normal composite construction in the draft code for buildings are 0·75 in simply supported beams and 0·7 in continuous beams.

With regard to composite reinforced concrete, there would appear to be two possible approaches to the problem. One would be to adopt the characteristic strengths of shear connectors as given by the standard push-off test recommended in the code, but then with new partial safety factors determined for the conditions applying to composite reinforced concrete. The other approach would be to determine new characteristic strengths of shear connectors using a new type of test more appropriate to conditions in a composite reinforced concrete beam, but then with the same partial safety factors as above.

It was the latter approach which was felt to be the more appropriate and at the University of Manchester two types of test have been investigated, the push-off test and the beam-type test.

Push-off test

The term 'push-off' has been adopted here to distinguish it from the standard 'push-out' test. The essential difference is that in the push-off test the specimen is asymmetrical. A typical specimen is illustrated in Fig. 3.6; it is intended to represent the web of a beam of composite reinforced concrete. The shear connector used in the majority of the tests was the welded headed stud, this having found almost universal acceptance for normal construction.

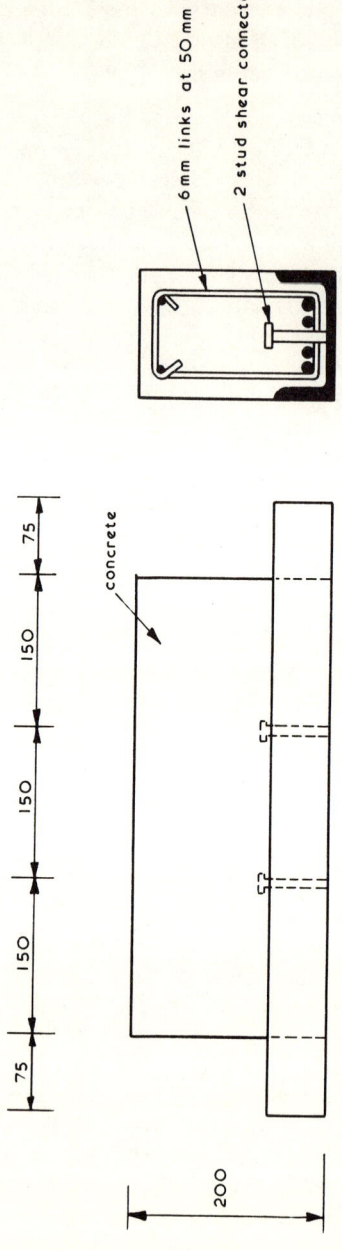

Fig. 3.6. Typical specimen used in push-off tests to determine strength of shear connectors

The main difficulty with this particular test is that the load causes a moment at the interface resulting in a tendency for vertical separation of the concrete and the steel channel, and hence tensile forces in the studs. Whilst such tensile forces do occur in the studs of an actual beam, the precise ratio of tension to shear in a stud is unknown.

In order to simulate conditions in a beam it seems necessary to use an applied load at an appreciable eccentricity, i.e. similar to the position of the compression zone of a beam. This will then cause the type of inclined cracking that can be observed in a beam. However it now becomes necessary to apply vertical forces in order to restrain excessive separation at the interface. If these restraining forces were absent, the conditions would be appreciably different to those in a beam and the studs would fail largely as a result of the bending moment rather than the shear force. Nevertheless the restraining forces are somewhat indeterminate and this aspect of the testing arrangement leaves much to be desired.

The usual arrangements for the tests at the University of Manchester are as in Fig. 3.7. Bolts (A) secure the steel plate (B) to the test frame. Horizontal resistance to movement of the concrete is reduced by placing steel rollers between the plate and the concrete block. At the beginning of the test the force in the bolts is quite small, the nuts to the bolts being finger tightened only. During the tests the forces in the bolts increase, but no attempt was made to measure this force. The arrangement was satisfactory in that in many of the tests inclined cracking did

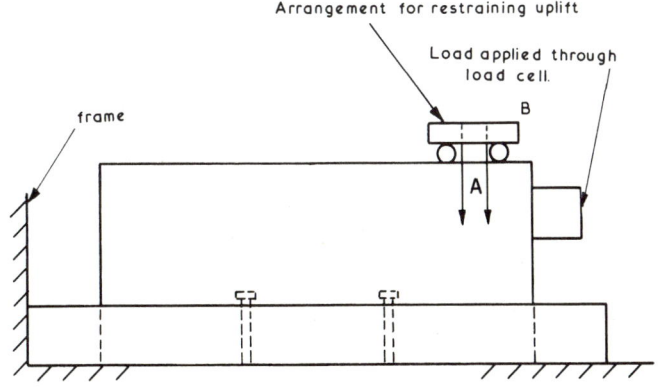

Fig. 3.7. Diagrammatic arrangement of the push-off tests

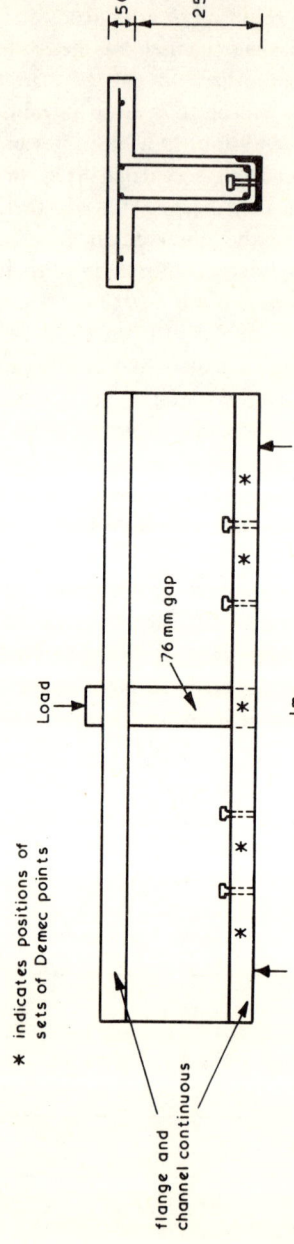

Fig. 3.8. Typical specimen used in the beam-type tests to determine the strength of shear connectors

occur, although the cracks were always much finer than those which occur in a beam.

In nearly all cases the specimens with studs tested in this way failed by shearing of the studs at the interface at loads slightly higher than the characteristic strengths quoted in the draft code. From this it can be inferred that this particular test was not in fact more stringent than the standard test for determining the shear strength of studs. Since it is possible that greater uplift forces would occur in studs in beams of composite reinforced concrete than does occur in the push-off tests, doubt remains as to whether the particular arrangement is sufficiently representative of actual behaviour.

Beam-type specimens

In an attempt to devise a standard test for determining the strength of shear connectors in which conditions are nearer to the actual conditions, a beam-type of push-off specimen was devised. The idea was obtained from a study of bond tests for reinforced concrete as described by Snowdon.[2]

The specimens were essentially short T-beams. Details of their dimensions and shear connection are shown in Fig. 3.8. There is a 75 mm gap in the concrete web at midspan, but the concrete is continuous in the top flange. The steel channel is continuous at the soffit of the beam. Thus at midspan the channel takes the whole of the tensile force whilst the flange of the beam takes all the compressive force.

The specimens were supported over a span of 1·2 m and load applied at midspan through a 75 mm wide plate bedded on the top face. Longitudinal strains in the channel were measured at midspan, the midpoints of the shear spans (i.e. between the two studs), and the points between the outer studs and the supports. Strains were also measured in the concrete flange at midspan.

As in the previously described push-off tests fine diagonal cracks occurred in the concrete webs, beginning from the positions of the inner studs. Associated with these diagonal cracks was a vertical flexural-type crack across the bottom of the flange at midspan. Collapse occurred as the result of the two studs at one end shearing simultaneously.

The force per stud causing shearing was calculated as the change in force in the channel between midspan and the section near the support divided by the two studs. The force in the channel at midspan was calculated in two ways: from the strain readings on the channel and from a knowledge of the bending moment at midspan and the lever arm of the internal forces. In general the two values were nearly identical.

The force in the channel at the section between the last stud and the support was calculated from the strain readings only. In theory, assuming transfer of force by shear connectors only, there should be zero force in the channel at this position, but the strain readings indicated otherwise. Presumably the frictional restraint between the channel and the concrete develops this force in the channel.

The results from these tests, therefore, tended to be slightly lower than those from the simple push-off tests, but nevertheless at values close to the characteristic strengths given in the draft code of practice.

The most likely reason for the slightly lower strengths in the case of the beam-type tests is that in calculating the values from the beam-type tests some allowance was made for the frictional forces between channel and concrete, whereas this was not the case in the push-off tests. The frictional restraint in the beam-type specimens will be greatest where the bearing stress between channel and concrete is the highest, and this will be in the region of the support. Thus it is felt that the method of calculation will have allowed for the majority of the frictional restraint. The light oiling of the surface of the channel prior to concreting is unlikely to have had much influence on the frictional resistance to sliding. In retrospect greasing of the channel would have been better for reducing the frictional resistance to a minimum.

Typical results for a beam-type specimen with 13 mm stud shear connectors and a concrete strength of 32 N/mm^2 are as follows.

Force in channel at midspan from bending moment
    calculation   = 129 kN
Force in channel at midspan from strain readings   = 128 kN
Force in channel near support   = 27 kN
Shear force in a stud at failure $= (129 - 27)/2$ = 52 kN

Design of stud shear connectors

It was stated earlier that, of the two possible approaches to the problem of the design strength of shear connectors in composite reinforced concrete, it was felt to be more appropriate to develop a new type of standard test which tended to reproduce more accurately the conditions in composite reinforced concrete. As a result of the tests carried out, doubt now exists as to whether this is a suitable approach. Although the inclined cracking occurring in the webs of beams was simulated in the push-off tests, the inclined cracks in the tests remained finer than those in beams near their ultimate load. Thus the uplift forces in the push-off and beam-type tests may be somewhat lower than can occur in a beam. Whilst more refined methods of push-off tests could improve this facet of behaviour, it seems doubtful that their development would be worthwhile.

It is currently considered that the alternative approach may be more appropriate, i.e. to adopt the standard characteristic strength of the shear connectors given in the code of practice but with factors suitable for composite reinforced concrete, these not necessarily being those used in normal composite construction.

It would seem likely that the best method to obtain suitable design factors would be to test full size beams of composite reinforced concrete with varying degrees of shear connectors. The majority of tests on full size beams so far carried out at the University of Manchester have aimed mainly at obtaining information on bending and shearing under conditions of 'adequate' shear connection. This adequate shear connection has in general been determined by using a few more stud shear connectors than would be required by the recommendations of CP 117 or more recently the new draft code. In none of the tests has failure been promoted by the shear failure of the connectors. In the future it would be desirable to test full size beams in which the stud shear connectors are more economically proportioned.

Until such time that more information is available it is proposed that for the design of all types of beams of composite reinforced concrete the design strength of the stud shear connectors shall be assumed as the characteristic strengths given in the draft code of practice for normal composite construction multiplied by 0·7, this being the factor in the draft code appropriate to continuous beams.

Reference has been made to the development and spreading of yielding of the steel channel at loads higher than the service load and Fig. 3.3 illustrates the region of full yield at the ultimate load of the beam designed earlier in chapter 1. This must be taken into account in the proportioning of the shear connectors. For example the whole of the maximum force in the steel channel must be transmitted by shear connectors contained within the outer regions of the beam (viz, regions AB and CD in the example in Fig. 3.3). Since at the load causing flexural failure the force along the channel in the central region BC remains constant, no shear connectors are required in this region for the transmission of forces between concrete and steel channel. Nevertheless some shear connectors will be necessary in this central region, partly to prevent vertical separation but also to provide the necessary transmission of forces between concrete and steel during the service load stages.

Design example

To illustrate the method of design of shear connectors the calculations will be presented for the proportioning of the connectors for the beam previously designed in chapter 1 and illustrated in Fig. 1.12. The type of shear connectors used are 19 × 100 mm studs.

Yield force in channel = 4552 × 0·25 × 0·93	= 1060 kN
Design strength of a stud = 0·7 × 100	= 70 kN
Number of studs required to develop yield force	= 15
Bending moment causing full yield of channel = 5108 × 0·25 × 0·572	= 730 kNm

Under the ultimate load this would occur at 2·5 m from the support. Thus 15 studs would need to be affixed in each of the outer 2·5 m lengths of beam. This entails a spacing of 165 mm.

In the central 4 m length of beam only 'nominal' studs at a spacing of, say, 300 mm are required. Nevertheless it is necessary to check that these nominal studs are satisfactory at the full service load, namely that the maximum force on a particular stud does not attain the design strength of that stud. For a constant spacing of studs in this central region the critical stud would be that positioned at the end of the region, i.e. at 2·5 m from the support. For T-beams it will be sufficiently accurate to calculate the change in force in the channel from one connector to the next using the change in bending moment.

Design load for serviceability (see Appendix 1)	= 60·6 kN/m
Bending moment at 2·5 m from support	= 494 kNm
Bending moment at 2·8 m from support	= 527 kNm
Increase in bending moment from one stud to next	= 33 kNm
Increase in steel force from one stud to next = 33/0·572	= 58 kN
Increase in channel force from one stud to next = 58 × 4552/5108	= 52 kN

Thus the force on the stud at the critical position of the central region is 52 kN which is less than the design strength of the stud.

Further information on what would be the maximum possible spacing of nominal studs in the central region of a beam is desirable.

It is important to note that the draft code of practice for normal composite construction stipulates that the diameter of a stud shear connector should not exceed twice the thickness of the plate to which it is welded. In this design example this stipulation was not quite satisfied — the diameter was 19 mm while twice the thickness of plate is 18·2 mm.

This is another area requiring further investigation. Is such a recommendation necessary for composite reinforced concrete? Certainly there seems no reason to differentiate between normal composite construction and composite reinforced concrete in this regard.

No information has been obtained at the University of Manchester regarding the critical relationship between the diameter of stud and the thickness of steel plate except that the code ruling has not always been adhered to in tests. This applies to both beam tests and push-off tests. In none of these cases was the shear strength of the shear connector adversely affected.

Should the ruling be found to be applicable to composite reinforced concrete, the diameter of studs used would sometimes need to be slightly smaller than the diameters usually adopted in normal composite construction.

ALTERNATIVE TYPES OF SHEAR CONNECTORS

The cost of stud shear connectors in normal composite construction is quite significant and this is clear from the example of the cost of materials in chapter 1 (see Table 1.3). Indeed it is this cost which has led to recent proposals for effecting some economy in normal composite construction by using fewer shear connectors than would be required for full interaction of the steel and concrete and accepting a lower ultimate resistance moment of the composite section.

One of the cost advantages of composite reinforced concrete is that it requires fewer shear connectors than the equivalent section of normal composite construction. This, of course, is due to the reduced cross-section of the channel in composite reinforced concrete compared with that of the I-beam in normal composite construction. It is found that the saving in the number of shear connectors is approximately 1/3

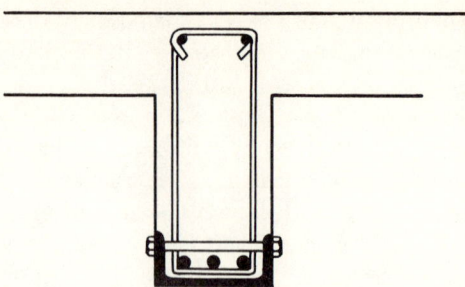

Fig. 3.9. Section through a beam having transverse bolt shear connectors

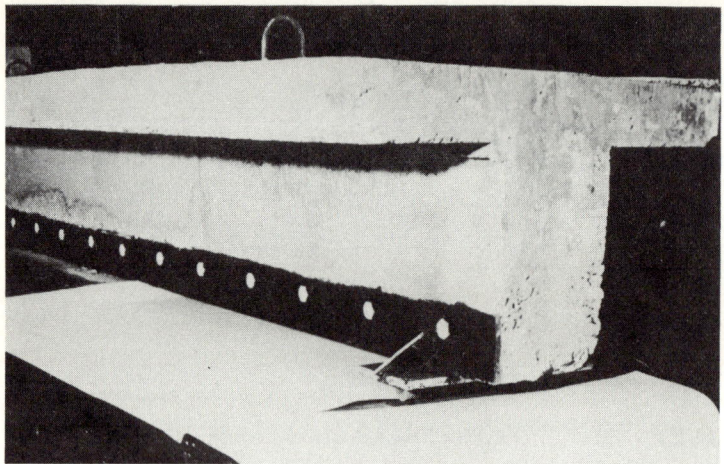

Fig. 3.10. Beam having transverse bolt shear connectors

Table 3.1. Results of push-off tests by Clarke and Nelson[3]

Diameter of bolt: mm	Over-size of holes: mm	Maximum load per shear connector: kN
12	0·4	69
12	1·6	72
12	2·4	68
16	1·6	110

Table 3.2. Results of push-off tests by Cunningham [4]

Diameter of bar: mm	Maximum load per shear connector: kN	Critical load per shear connector: kN
10	55	29
12	88	42
16	114	57

of the number in the corresponding beam of normal composite construction.

Transverse bolt

Despite this immediate advantage, investigations have proceeded in an attempt to reduce the cost even further by seeking alternatives to the expensive stud. One alternative is the transverse bolt (Fig. 3.9). This passes through holes in the flanges of the channel, being positioned subsequent to the positioning of the reinforcement cage, but prior to the concreting. A beam with such shear connectors is shown in Fig. 3.10. The advantages of such an arrangement are that

(a) the bolt costs slightly less than the headed stud
(b) no expensive stud welding equipment is required
(c) the bolt is in double shear as opposed to the single shear in the cantilever stud.

The main disadvantage is that the holes in the channel have to be drilled (or punched) prior to delivery to site.

To ascertain the strength of such bolts, push-off tests were carried out by Clark and Nelson[3] and the results of their tests are summarized in Table 3.1. The values of the maximum load are the averages from two push-off tests. The 28 day cube strengths of the concrete ranged from 27 to 33 N/mm^2.

One of the variables in the tests was the over-size of the holes in which the bolts fitted, but this has had no apparent effect on the strength. In the specimens incorporating 16 mm bolts failure occurred as a result of crushing of the concrete at the position of the applied load and shearing of the bolts did not occur. In all the specimens using 12 mm bolts shearing of the bolts occurred always at the position of a thread. In all cases the failure loads were appreciably higher than the

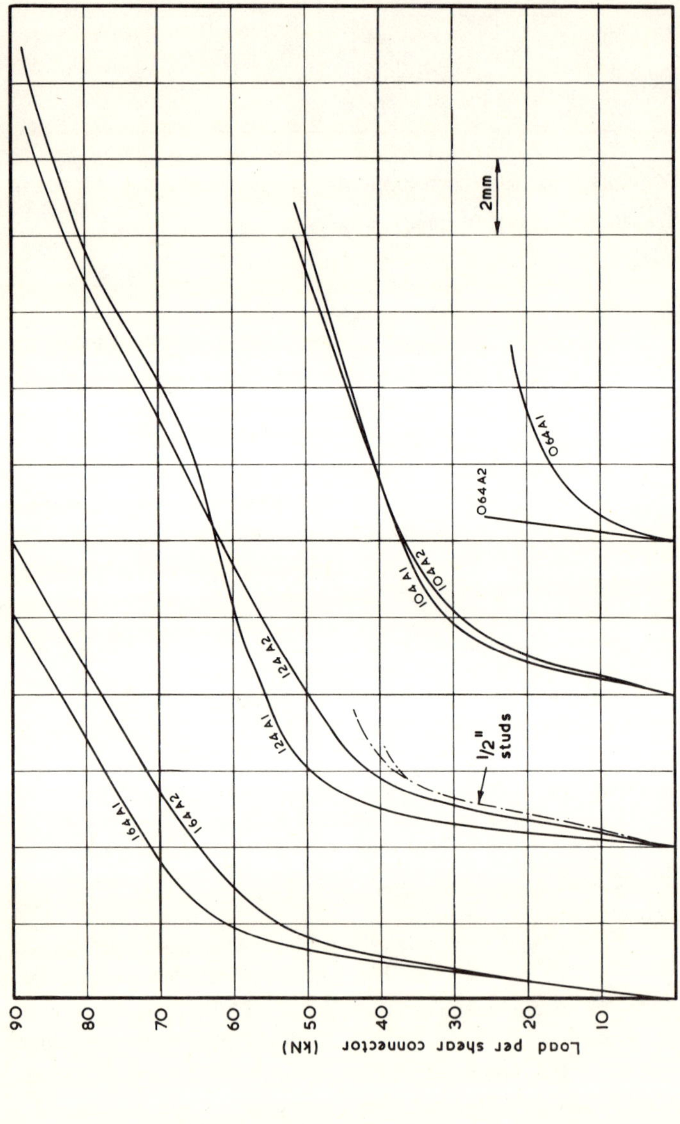

Fig. 3.11. Typical load–slip curves for push-off specimens with transverse bar shear connectors

characteristic strengths of the corresponding stud, but certainly not twice these values.

Transverse bar

Another possible shear connector is the transverse bar, i.e. a plain bar placed through holes in the channel. Compared with the bolt, the bar is significantly cheaper. Moreover there are no threads to affect the strength. Its disadvantage is that, until the concrete has hardened, the bar is not fixed and during construction on site the bar could be displaced. Results from push-off tests by Cunningham[4] are given in Table 3.2.

The values are the averages of four tests. The over-size of holes again were varied but this had no apparent effect on the failure load. Although shearing of the bars occurred in all specimens having 10 and 12 mm dia. bars, the specimens having 16 mm dia. bars failed as a result of crushing of the concrete at the position of the applied load. The 28 day cube strength of the concrete ranged from 34 N/mm² to 49 N/mm² with an average value of 42 N/mm².

For the size where both bars and bolts failed as a result of shearing, the bars, as would be expected, sustained the higher average load, this being nearly twice the characteristic strength of the studs of the same diameter.

Unfortunately the very high strengths of the transverse bar shear connectors cannot be used for proportioning shear connectors in beams because of the very high slips associated with the high load. Typical load—slip curves are shown in Fig. 3.11 where it will be seen that slips of 20 mm were not unusual at the maximum load. Such slips, if occurring in a beam, would allow such loss of interaction that the failure load would be considerably lower than the theoretical failure load calculated assuming full interaction.

From a study of the load—slip curves, in the majority of cases the change in the rate of slip occurs at the load causing a slip of approximately 2 mm. Subsequent to this slip the rate of slip increased considerably and during these load stages the bars could be seen being drawn into the holes. An examination of test data from standard push-out tests using studs shows that at a slip of 2 mm the stud has reached a substantial proportion of its ultimate strength. It would therefore seem that, if such slips could be accommodated in proportioning studs in beams, the loads corresponding to a slip of 2 mm could safely be equated to characteristic strength, at least to facilitate an initial comparison of costs. Such loads have been termed the 'critical loads' and have been included in Table 3.2.

Using the critical loads as the criterion for design, two full-sized beams with transverse bar shear connectors have been tested (Fig. 3.12). They behaved satisfactorily as regards both the ultimate load (failing in flexure) and the load—slip characteristics. None of the bar shear connectors failed or showed any signs of distress.

A comparison of the critical loads of 12 mm and 16 mm bars with the characteristic strengths of the correspondingly sized studs suggests that the use of bars would entail using approximately 4/3 times the number of studs of the same diameter. The use of bars would therefore appear economically advantageous, but further thought on this is desirable. Certainly more experimental work on full sized beams with bar shear connectors would be required before they could be used in practice with confidence.

LONGITUDINAL SHEARING

The draft code of practice for normal composite construction draws attention to the possibility of longitudinal shear failure occurring outside the shear connectors. For example longitudinal shearing can occur

Fig. 3.12. Beam having transverse bar shear connectors; (the advantage of the bar shear connector is its cheapness; its disadvantage is that it could be dislodged during construction, although this could probably be overcome)

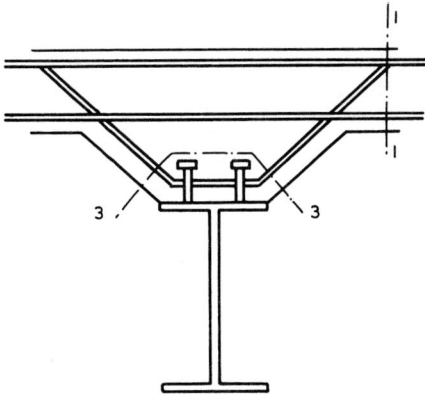

Fig. 3.13. Planes for possible longitudinal shearing in normal composite construction

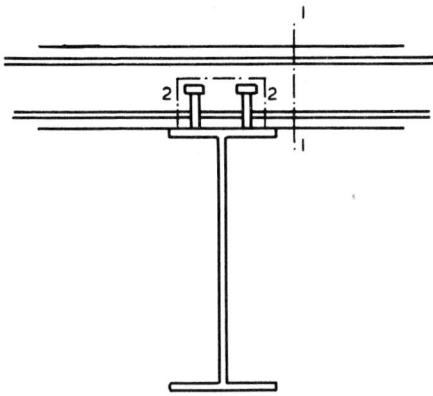

Fig. 3.14. Planes for possible longitudinal shearing in deep haunched construction

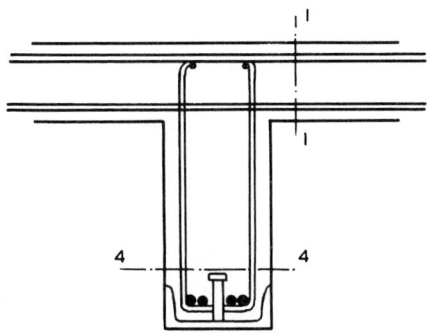

Fig. 3.15. Planes for possible longitudinal shearing in composite reinforced concrete

in normal composite construction along the planes illustrated in Fig. 3.13, or in haunched composite construction along planes illustrated in Fig. 3.14. The code gives expressions for ensuring that such shear failures do not occur, for example the longitudinal force Q per unit length should satisfy

$$Q < 0.9\, sL_s + 0.7\, A_e f_{ry}$$

where L_s is the length of the shear plane, s is a stress of 1 N/mm^2, and A_e is the effective area of reinforcement through the shear planes.

Clearly for composite reinforced concrete similar consideration should be given to plane 4 in Fig. 3.15. The question that arises is to what extent the vastly different conditions (e.g. in a tension zone) affect the shear strength and to what extent the above expression would need to be modified. It is a complicated problem and no work on this has so far been pursued. It should be added though that in some of our tests the above condition has not been satisfied but with no apparent detrimental effects.

SUMMARY

Regarding the design of the shear reinforcement, tests on beams indicate that the proportioning of shear links in composite reinforced concrete in accordance with the recommendations of CP 110 would be more than really required from the strength standpoint. However, such amounts would probably be necessary to satisfy the requirements of limited crack width at the service load. To clarify the issue, further tests on practical sized beams are required.

On the design of stud shear connectors, it is suggested that the same characteristic strength used in the design of normal composite construction could be adopted for composite reinforced concrete but with a safety factor of 0.7. This factor is slightly lower than that used for simply supported beams in normal composite construction. A further difference in the two types of construction is that in composite reinforced concrete the shear connectors would need to be concentrated in those outer regions of the beams where the steel channels do not attain their full yield stress at the ultimate load.

Although exploratory tests indicate that the transverse bar and bolt shear connectors are technically satisfactory and more economical, further full scale beam tests are needed before they could be used with confidence.

REFERENCES

1. TAYLOR R., MILLS P. E. and RANKIN R. I. Tests on concrete beams with mixed types of reinforcement. *Mag. Concr. Res.*, 1978, Vol. 30, June, No. 103, 73–88.
2. SNOWDON L. C. 'Classifying reinforcing bars for bond strength.' Building Research Establishment, Garston, 1970. Current paper 36/70.
3. TAYLOR R., CLARK D. S. E. and NELSON J. H. Tests on a new type of shear connector for composite reinforced concrete. *Proc. Instn Civ. Engrs*, Part 2, 1974, Vol. 57, Mar., 177.
4. TAYLOR R. and CUNNINGHAM P. Tests on transverse bar shear connectors for composite reinforced concrete. *Proc. Instn Civ. Engrs*, Part 2, 1977, Vol. 63, Dec., 913–920.

CHAPTER 4

Miscellaneous problems

PROBLEM OF DEFLEXION

It has been demonstrated that beams of composite reinforced concrete behave essentially the same as beams of reinforced concrete, and this applies also to the deflexion characteristics. With regard to deflexions up to the service load it is found that the method for calculating deflexions given in CP 110 can be applied to beams of composite reinforced concrete with a reasonable degree of accuracy, although some adjustment may be necessary in order to obtain suitable span/depth ratios for use in design.

The tables for establishing span/depth ratios for normal reinforced concrete given in CP 110 can be applied directly only to rectangular beams. Span/depth ratios for T-beams have been related to those of rectangular beams by factors which depend on the ratios of flange and web widths. As a result, in contrast to the 'precise' span/depth ratios for rectangular reinforced concrete beams, those for T-beams are 'approximate'. Since beams of composite reinforced concrete will always be of the T-beam type, it is perhaps preferable to provide new span/depth ratios which can be applied directly to T-beams rather than indirectly and approximately via the ratios for rectangular beams.

The essential difference between rectangular beams and T-beams from the point of view of deflexion is the degree of tension stiffening offered by the concrete in the region of the tension steel. In T-beams there is relatively less concrete at the level of the tension steel and so there is less tension stiffening and therefore somewhat higher deflexions. A beam of composite reinforced concrete is in effect a highly reinforced T-beam so that there is relatively little tension stiffening and the effect can be ignored in calculations.

Since 'effective depth' has a less precise meaning for beams of composite reinforced concrete, it would be more convenient for span/depth ratios to be related to 'overall depth'.

Method of calculation

For beams in which the moment—curvature relationship is linear, the maximum deflexion of a beam can be expressed in the form

$$v_{max} = k \, \phi_{max} \, L^2$$

where k is a constant depending on the type of loading and the end conditions of the beam (e.g. $k = 5/48$ for a simply supported beam subjected to a uniformly distributed load), and ϕ_{max} is the maximum local curvature.

The curvature can be expressed in terms of strains at the top and bottom of the beam, for example

$$\phi = \frac{\epsilon_t + \epsilon_b}{D}$$

where D is the overall depth of the beam.

Alternatively the curvature can be expressed in terms of strains at the top and the level of the reinforcement, for example

$$\phi = \frac{\epsilon_t + \epsilon_r}{d}$$

where d is the effective depth of the beam.

It is the latter expression for curvature which is used for reinforced concrete calculations, but it is the former which is suggested as being more appropriate for beams of composite reinforced concrete.

In normal circumstances in reinforced concrete the strains used in the calculation are the average strains at the position of maximum curvature after tension stiffening has been allowed for, but, as already explained, for the T-beams of composite reinforced concrete, this effect can be ignored. Thus

$$v_{max} = k \, (\epsilon_t + \epsilon_b)_{max} \, L^2/D$$

It remains to compare the values obtained using this equation with experimental data.

Basis of span/depth ratios for use in design

Re-arranging the last equation gives the span/depth ratio of a beam as

$$\frac{L}{D} = \frac{v_{max}}{L} \frac{1}{k} \frac{1}{(\epsilon_t + \epsilon_b)_{max}}$$

In this expression the ratio v_{max}/L is the limiting value of the sag of a beam. The sag specified in CP 110 is 1/250 but the code allows this to be modified to take into account a number of possible advantageous effects, some of which will always occur in a beam of a real structure. The possible effects are

(a) actual strength of concrete $>$ assumed
(b) actual creep and shrinkage $<$ assumed
(c) actual loads $<$ assumed
(d) partitions and finishes increase stiffness
(e) rotational restraints at the supports
(f) restraint to longitudinal expansion.

It is not necessary to explain here precisely how these possible effects were allowed for in calculating the span/depth ratios for reinforced concrete; suffice it to state that the values of v_{max}/L used in those calculations were

$$\frac{v_{max}}{L} = \frac{1}{183}$$

for simply supported beams, and

$$\frac{v_{max}}{L} = \frac{1}{210}$$

for continuous beams.

These values used for normal reinforced concrete are also appropriate for composite reinforced concrete. Thus, for example, the resulting equation for estimating permissible span/overall depth ratios for simply supported composite reinforced concrete beams would be

$$\frac{L}{D} = \frac{1}{183} \frac{48}{5} \frac{1}{(\epsilon_t + \epsilon_b)_{max}} = \frac{52 \times 10^{-3}}{(\epsilon_t + \epsilon_b)_{max}}$$

This assumes that the method of calculation for reinforced concrete can also be used for composite reinforced concrete. It has yet to be determined from a study of experimental data whether any modification to the method of calculation is required.

The values of $(\epsilon_t + \epsilon_b)_{max}$ depend on

(a) the stress level in the channel at the bottom of the beam at the full service load (this determines ϵ_b)

(b) the total area of tension steel (this determines the depth of the neutral axis and hence the value of ϵ_t)

(c) the shape of the beam (particularly the ratio of depth of flange to overall depth since this will influence the depth of the neutral axis).

It should therefore be possible to obtain tables comparable to those in CP 110 but for which span/depth ratios can be obtained for T-beams directly. No attempt has so far been made to provide such tables and there is scope here for some original thought on the subject.

W_s is the full service load

W_p is the permanent load $(=\tfrac{3}{4}\,W_s)$

$\triangle_i$ is the instantaneous deflexion
        due to permanent load.

$\triangle_p - \triangle_i$ is the additional deflexion
        due to the long term effects.

$\triangle_s - \triangle_p$ is the extra instantaneous
        deflexion as the load increases to W_s

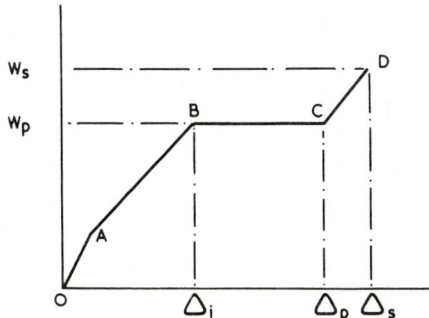

Fig. 4.1. Assumed load–deflexion characteristics for calculating service load deflexions and span/depth ratios

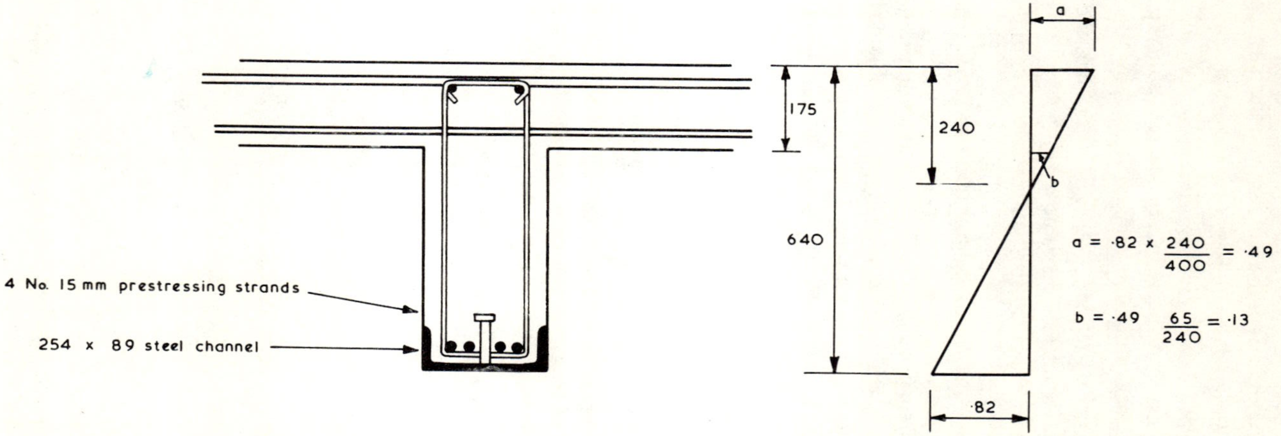
4 No. 15 mm prestressing strands
254 x 89 steel channel
175
640
240
a
b
·82
a = ·82 x 240/400 = ·49
b = ·49 65/240 = ·13

Fig. 4.2. Cross-section of beam used in example 1 to calculate deflexion with the assumed strain diagram used in the calculation

Examples of calculation of deflexion

The principles adopted here are exactly as those adopted in the calculation of span/depth ratios for normal reinforced concrete with the exception that tension stiffening by the concrete is ignored.

The assumed load—deflexion diagram is as shown in Fig. 4.1. The long term effects of creep and shrinkage are indicated by the horizontal part BC. This occurs under the permanent load which is assumed as 3/4 of the full service load. The maximum deflexion is therefore made up of the permanent deflexion Δ_p (which includes the long term effects) and the short term additional deflexion (represented by part CD) as the load is increased from W_p to W_s.

It will be assumed in the calculation that there is essentially no change in stress in the channel due to the long term effects. This is a reasonable assumption as the lever arm between the centroids of compression and tension in such T-beams must remain virtually unaltered.

EXAMPLE 1

The beam to be analysed is the first beam designed in chapter 1. Its cross-section is shown in Fig. 4.2.

Short term E value of concrete (from CP 110)	$= 28 \text{ kN/mm}^2$
Long term E value of concrete (from handbook)	$= 7 \cdot 5 \text{ kN/mm}^2$
Total area of tension steel	$= 5108 \text{ mm}^2$
Effective width of flange	$= 2 \cdot 05 \text{ m}$
Depth of centroid of steel	$= 614 \text{ mm}$
Percentage of steel	$= 0 \cdot 41\%$

Maximum strains under the permanent load

The maximum stress at the soffit of the beam calculated in chapter 1 under the full service load was 220 N/mm^2.

Maximum stress at soffit under permanent load
$= 3/4 \times 220$ $= 165 \text{ N/mm}^2$
thus ϵ_b $= 0 \cdot 82 \times 10^{-3}$

As the neutral axis under the long term permanent load is probably below the flange, a trial and error procedure will be used to establish its position. Assume a depth of 240 mm. The strain diagram would then be as shown in Fig. 4.2.

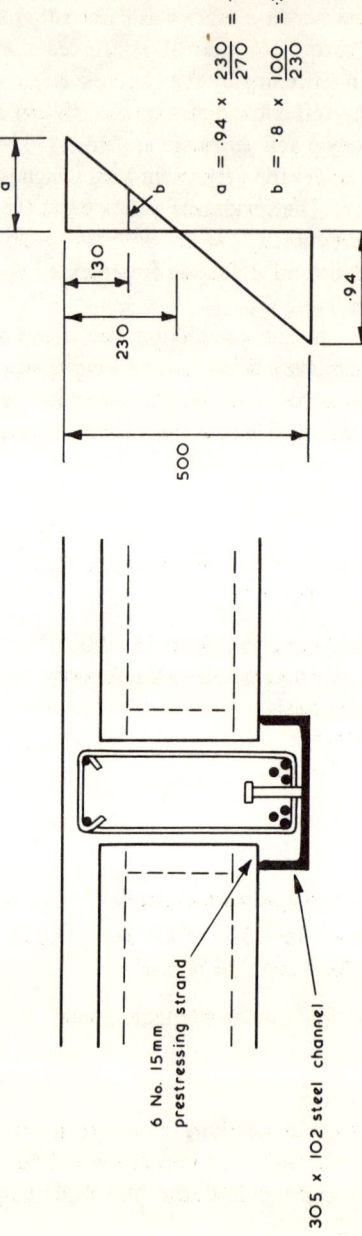

Fig. 4.3. Cross-section of beam used in example 2 to calculate deflexion with the assumed strain diagram used in the calculation

Mean strain in the concrete in the flange	$= 0.31 \times 10^{-3}$
Corresponding mean stress in the concrete	$= 2.36 \text{ N/mm}^2$
Compressive force in the flange	$= 845 \text{ kN}$
Tensile force in steel $= \frac{3}{4} \times 1070$	$= 800 \text{ kN}$

The discrepancy between the calculated tensile and compressive forces means that the assumed depth of the neutral axis is slightly in error. However the agreement is near enough to be able to correct the assumed concrete strains by direct proportion.

Thus $\epsilon_t = 0.49 \times 10^{-3} \times 800/845$	$= 0.465 \times 10^{-3}$
and $(\epsilon_t + \epsilon_b)_{max}$ for the permanent load	$= 1.29 \times 10^{-3}$

Additional strains as the load is increased to full service load

The neutral axis for this short term loading will lie within the depth of the slab. Hence the depth of the neutral axis can be calculated from the simple equation appropriate to rectangular beams.

Depth of neutral axis	$= 130 \text{ mm}$
Additional strain at bottom of beam	
$= 1/3 \times 0.81 \times 10^{-3}$	$= 0.27 \times 10^{-3}$
Additional strain at top of beam	
$= 0.27 \times 10^{-3} \times 130/510$	$= 0.07 \times 10^{-3}$
Thus $(\epsilon_t + \epsilon_b)_{max}$ for additional load	$= 0.34 \times 10^{-3}$

Effect of shrinkage

It has been assumed that Table 63 of CP 110 is applicable to composite reinforced concrete.

Thus $(\epsilon_t + \epsilon_b)_{max}$ for shrinkage	
$= 0.52 \times 300 \times 10^{-6}$	$= 0.156 \times 10^{-3}$
Hence total value of $(\epsilon_t + \epsilon_b)_{max}$	$= 1.78 \times 10^{-3}$
and $v_{max} = \dfrac{5}{48} \times 1.78 \times 10^{-3} \times \dfrac{(9 \times 10^3)^2}{640}$	$= 24 \text{ mm}$

Maximum permissible calculated value of	
deflexion $= \dfrac{9 \times 10^3}{183}$	$= 49 \text{ mm}$

This comparison shows that from the deflexion standpoint the beam could have been much shallower. A shallower beam will therefore be analysed in the next example.

EXAMPLE 2

The beam to be analysed is the second beam designed in chapter 1. Its cross-section is shown in Fig. 4.3.

Total area of tension steel	$= 6715 \text{ mm}^2$
Effective width of flange	$= 2 \cdot 1 \text{ m}$
Depth of centroid of steel	$= 471 \text{ mm}$
Percentage of steel	$= 0 \cdot 68\%$

Maximum strains under the permanent load

The maximum stress at the soffit of the beam calculated in chapter 1 under the full service load was 250 N/mm².

Maximum stress at soffit $= \frac{3}{4} \times 250$	$= 188 \text{ N/mm}^2$
Thus ϵ_b	$= 0 \cdot 94 \times 10^{-3}$

Again, as the neutral axis under the permanent load is probably below the solid part of the flange (i.e. within the hollow part of the precast units), a trial and error procedure will be used to establish its position. Assume a depth of 230 mm. (Note that the depth of the solid part of the slab is taken as 130 mm). The corresponding trial strain diagram is given in Fig. 4.3.

Mean strain in the concrete in the flange	$= 0 \cdot 57 \times 10^{-3}$
Corresponding mean stress	$= 4 \cdot 3 \text{ N/mm}^2$
Compressive force in the flange	$= 1170 \text{ kN}$
Tensile force in the steel $= \frac{3}{4} \times 1630$	$= 1220 \text{ kN}$

Again there is near enough correspondence for the assumed concrete strains to be corrected by direct proportion.

Thus $\epsilon_t = 0 \cdot 8 \times 10^{-3} \times 1220/1170$	$= 0 \cdot 84 \times 10^{-3}$
and $(\epsilon_t + \epsilon_b)_{max}$ for the permanent load	$= 1 \cdot 78 \times 10^{-3}$

Additional strains as the load is increased to full service load

Depth of neutral axis	$= 123 \text{ mm}$
Additional strain at bottom of beam $= 0 \cdot 94 \times 10^3/3$	$= 0 \cdot 31 \times 10^{-3}$
Additional strain at top of beam $= 0 \cdot 31 \times 10^{-3} \times 123/377$	$= 0 \cdot 10 \times 10^{-3}$
Thus $(\epsilon_t + \epsilon_b)_{max}$ for additional load	$= 0 \cdot 41 \times 10^{-3}$

Effect of shrinkage

$(\epsilon_t + \epsilon_b)_{max}$ for shrinkage $= 0.62 \times 300 \times 10^{-6}$ $= 0.19 \times 10^{-3}$

Hence total value of $(\epsilon_t + \epsilon_b)_{max}$ $= 2.38 \times 10^{-3}$

and $v_{max} = \dfrac{5}{48} \times 2.38 \times 10^{-3} \times \dfrac{(9 \times 10^3)^2}{500}$ $= 40$ mm

Again this is less than the maximum permissible value of 49 mm.

From CP 110 the permissible span/effective depth for a reinforced concrete T-beam having a reinforcement percentage of 0.68 and a service stress of 238 N/mm² is

$$20 \times 1.13 \times 0.8 = 18$$

The actual span/effective depth for the beam in question is

$$9/0.471 = 19$$

Thus on the evidence of the span/depth ratios the beams would be slightly 'unsafe' from the deflexion standpoint, whereas the calculation of curvatures shows the beam to be slightly on the safe side. There are two reasons for this slight discrepancy. One is that the shrinkage curvatures in these calculations adopted the values recommended in CP 110 whereas the values adopted for computing the CP 110 span/depth ratios were somewhat higher. The second reason is that the span/depth ratios in CP 110 are only approximate for T-beams and for these particular T-beams the error gives values slightly on the 'safe' side.

Comparison with experimental data

The load—deflexion curves for the beams reinforced with prestressing strand tested by Rankin (reference 3 of chapter 1) are shown in Fig. 4.4 for the range of loading up to the service load. The service load has been assumed to be that found by dividing the theoretical ultimate load by 1.75. The theoretical values of the midspan deflexion at the service load are indicated on the curves. It will be noted that in all four cases the experimental values exceed the theoretical values, the ratios of the experimental to theoretical deflexions being in the range 1.12—1.25. For reinforced concrete beams the ratios would be expected to be within the range 0.85—1.15.

Thus the indications from these tests are that the deflexions of beams of composite reinforced concrete are slightly higher than the method of calculation based on reinforced concrete theory would predict. If on

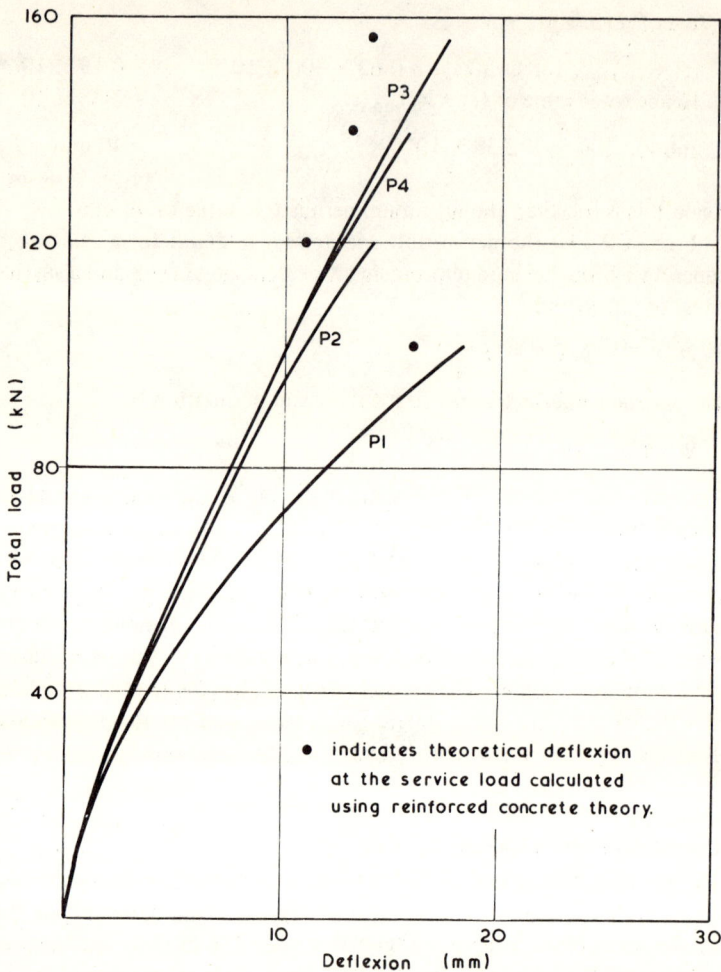

Fig. 4.4. Load–deflexion curves of Rankin's beams up to their service load

further investigation this does indeed prove to be the case it is most likely due to the difference in bond between the channel and the concrete on the one hand and reinforcing bars and the concrete on the other. However the tests are too few to draw firm conclusions at this stage. It is indeed possible that other aspects of the beams have influenced the deflexion characteristics. For example Rankin's beams had less shear reinforcement than that required by CP 110. Proportioning of the shear reinforcement more in accordance with the requirements of CP 110 may have reduced the deflexion at the service load. Moreover it might also be pointed out that few ordinary reinforced concrete T-beams have been tested and analysed with regard to deflexion. It is quite possible that such beams would also give a higher range of ratios of experimental to theoretical deflexions than 0·85—1·15 which is a range established largely from the analysis of reinforced concrete rectangular beams.

Clearly then more tests are necessary before any definite conclusions can be reached with regard to the method of calculating the deflexions of beams of composite reinforced concrete. Such tests should include long term effects. Not until such data is available will it be possible to translate the earlier equation into span/depth ratios suitable for design.

CRACKS AT THE SERVICE LOAD

Sagging moment regions

For the type of construction for which composite reinforced concrete was developed, namely where precast units or profiled steel sheeting are incorporated, the cracks which occur in the web of the composite reinforced concrete beam cannot be seen. Nevertheless it is of interest to record the maximum widths of cracks measured on beams tested in the laboratory.

Reference has been made in chapter 1 (reference 1) to the early tests by Burdon using Bristrand. In those tests the flexural cracks at the service load remained quite fine as illustrated in Fig. 4.5. Similarly the widths of the flexural cracks in the beams tested by Rankin are summarized in Fig. 4.6. The maximum width of the cracks in the beams reinforced with prestressing strand was 0·15 mm.

It must of course be remembered that these were short term tests and the widths would be expected to increase under long term loading. For the type of loading regime adopted for deflexion serviceability criteria (as in Fig. 4.1), it would be expected that the widest cracks at the full service load might attain maximum values of 0·3 mm, which is the maximum value permitted by CP 110. Needless to say further data on crack widths are required, including data on the effects of long term loading.

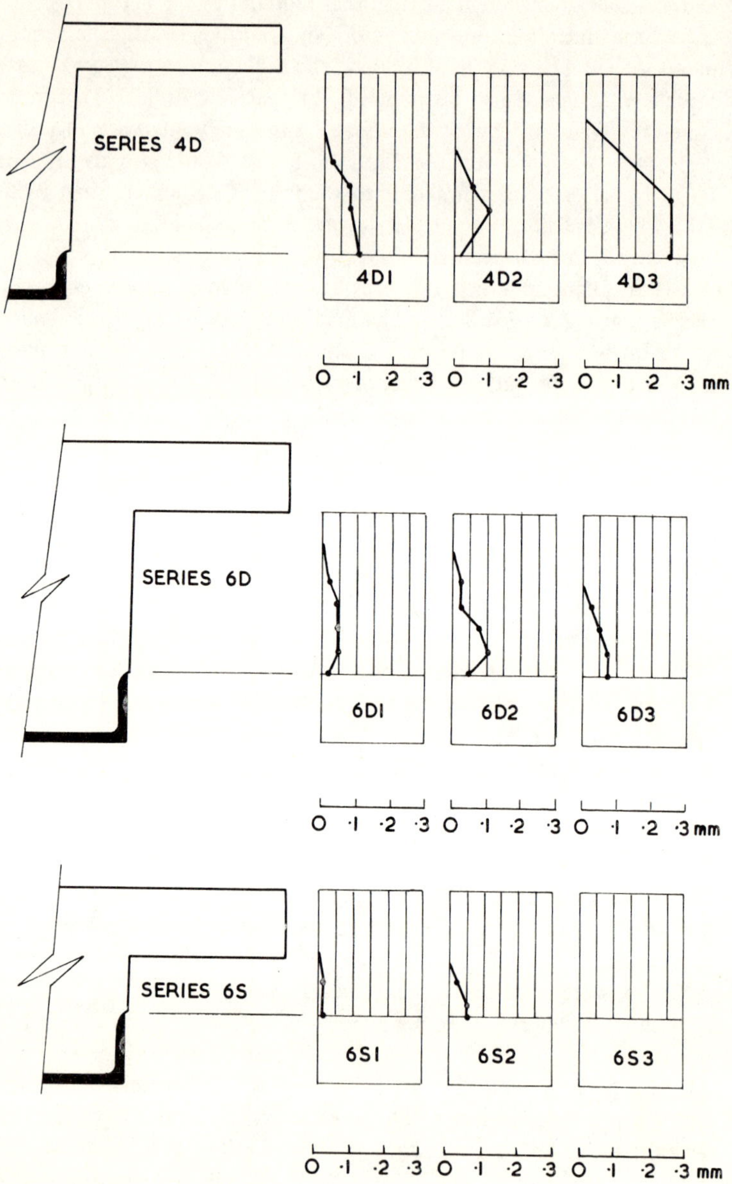

Fig. 4.5. Measured widths of flexural cracks at the service load in the beams with Bristrand tested by Burdon

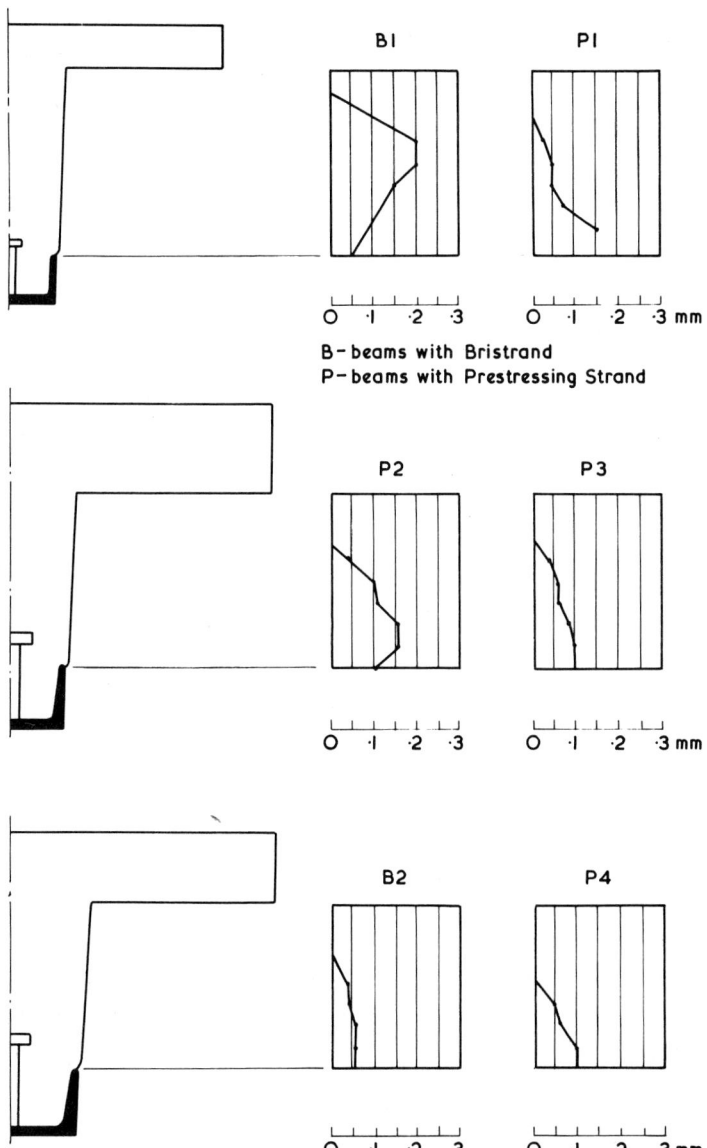

Fig. 4.6. Measured widths of flexural cracks at the service load in the tests by Rankin

Hogging moment regions

The cracking of beams of composite reinforced concrete in hogging moment regions is, of course, similar to that in ordinary reinforced concrete beams. One point of difference relates to the probability of being able to utilize plastic design for composite reinforced concrete in order to minimize the hogging resistance moment corresponding to the ultimate load. Whilst this is economically advantageous in the design of continuous beams (in that it allows a greater proportion of strand reinforcement to resist sagging bending moments as explained in chapter 2), it does make for more stringent conditions with regard to cracking in the region of the hogging moments at the service load, since the stresses in the reinforcement will be a high proportion of the yield stress. For example, in chapter 2 it was proposed that continuous beams be designed by assuming an ultimate hogging resistance moment equal to 1·25 times the service load hogging bending moment. This would mean that the stresses in the reinforcement at the full service load could be as high as 340 N/mm² for high yield reinforcing bars.

Whilst such high tensile stresses in normal reinforced concrete are not unknown, it remains to verify experimentally whether such stresses in composite reinforced concrete would lead to unacceptable wide cracking of the top slab. At the time of writing, the initial tests by Najmi (reference 2 of chapter 2) indicate that the cracks in the top slab at bending moments corresponding to the service load are quite small and well within the limits of acceptability. Again, though, more information is needed.

PROBLEM OF FATIGUE

The welding of stud shear connectors to steel at positions subsequently subjected to a high and variable stress is detrimental to the fatigue life of the steel. In beams of composite reinforced concrete tested under repeated loading, fracture of the channel occurs at a position of a stud. Only two such beams have been tested. The range of the load in these tests was from a maximum of the full service load to 1/3 of this value. Fracture of the channel occurred in both beams after approximately ½ million repetitions of load. Subsequent breaking open of the beams indicated that several studs adjacent to the fracture had sheared off, but it was not clear whether this occurred prior to, or as a result of, the fracture of the channel.

It is therefore clear that composite reinforced concrete is not a suitable material to use under conditions of repeated loading where large ranges of load can occur, e.g. in bridge structures. It should, however, be acceptable for the multi-storey office block type of structure where, not only is the range of loads normally a small part of the total service load, but the normal maximum load is appreciably less than the full design load.

The use of the alternative shear connectors, namely the transverse bolts or bars, does not improve the fatigue life significantly. Two beams, tested under conditions similar to those described above but using transverse bolt shear connectors and a maximum stress in the channel of 85% of the yield stress, failed due to fracture of the channel after approximately one million repetitions of load. Although welding was not involved here, the stress concentrations due to the shear connectors and the holes in the flanges of the channel produced the conditions to cause fatigue failure. During the tests fatigue cracks could be observed emanating from the holes prior to the actual fracture.

This would therefore seem to be a field where no further research is required. It must be accepted that composite reinforced concrete has a poor fatigue resistance and could not be used in structures subjected to heavy repeated loading.

PROBLEM OF FIRE RESISTANCE

The fire resistance of multi-storey structures plays a very important role in design, so a new form of construction cannot be proposed without referring to it, although at this stage no fire tests have yet been carried out.

The fire resistance of composite reinforced concrete will lie somewhere between the poor resistance of normal composite construction and the good resistance of traditional reinforced concrete. In comparison with composite construction the new material has the advantage that some of the steel is protected within the concrete so that the rise of temperature of this steel would be slower. It therefore seems possible that for the lower periods of fire resistance, say half an hour or one hour, no fire protection would be required. Where fire protection is required there is no doubt that the sprayed-on or painted-on systems will be cheaper for composite reinforced concrete than for composite construction because of the much smaller surface area of the steel section. The steel channel of composite reinforced concrete would be retained in position by the shear connectors and the concrete web and would not buckle in a fire as easily as does a steel I-beam. Thus, subsequent to the fire, if collapse has not occurred, the channel would regain most of its original strength.

In comparison with the fire resistance of traditional reinforced concrete, composite reinforced concrete is certainly at a disadvantage since all the steel in the former is protected by the concrete to some extent. Nevertheless one point of difference may be of interest. Under certain conditions the concrete cover in reinforced concrete can spall away leaving the reinforcing bars largely unprotected. In composite reinforced concrete the channel is retained in position by the shear connectors and this in turn retains the concrete in position. Thus the internal reinforcement will always have some protection.

Notwithstanding, for the multi-storey structure for which composite reinforced concrete is considered most suitable, it is anticipated that some protection of the steel channel will be essential.

SUMMARY AND CONCLUSIONS

This monograph summarizes the work carried out on composite reinforced concrete over a period of years at the Simon Engineering Laboratories of the University of Manchester. From the discussion it will be clear that there are areas in which further research is desirable. Indeed it was one of the purposes of the seminar to widen the research interest in the new material. Nevertheless it is felt that the stage has been reached whereby composite reinforced concrete could be used with confidence in actual construction despite the research continuing. After all, research in normal reinforced concrete still continues some 80 years after its initial use.

In most innovations the first move from a controlled laboratory situation to actual construction on site requires courage and something of an act of faith. However, in the case of composite reinforced concrete, whilst it is a completely new method of construction, it is not so far removed from the traditional forms of reinforced concrete and composite construction. The same fundamental principles apply. Tests have already shown that where design is in accordance with these principles, structures satisfactory from the technical standpoint will result.

One question remains: Are the economic advantages sufficient to warrant a move away from known and tried methods of construction? The Author believes that they are for certain types of structures and that composite reinforced concrete will be most advantageous for the construction of multi-storey structures where it is desired to use precast flooring in conjunction with in situ concrete frames for which the beams must be of a minimum depth.

There are several main advantages.

(a) The steel for resisting the longitudinal tensile forces is very compact (thus facilitating steel fixing).

(b) A small depth of construction can thereby be achieved (in contrast to reinforced concrete where such depths would produce a congestion of reinforcing bars).

(c) The precast flooring units are supported on the beam 'reinforcement' during construction resulting in a very small depth of beam beneath the precast units (in contrast to composite construction when precast units are used in conjunction with steel I-beams).

(d) No formwork is required for the in situ concrete beams.

(e) There is improved ability to resist hogging bending moments (in comparison with the traditional concrete materials).

(f) There is improved plasticity at the ultimate load allowing plastic design to be adopted (in contrast to traditional reinforced concrete whereby cumbersome trial and error methods for calculating moment redistribution are used).

APPENDIX 1

Calculations for the comparative designs in chapter 1

DESIGNS A (see Fig. 1.14)

Loading from a 1 m wide strip of slab

Slab self-weight $0.175 \times 23.6 \times 6$	= 24.8 kN
Finishes and partitions 1.5×6	= 9 kN
Characteristic dead load	= 33.8 kN
Characteristic live load 4×6	= 24 kN

Composite reinforced concrete

Dead load from slab	= 33.8 kN/m
Dead load of web of beam	= 2.8 kN/m
Characteristic dead load	= 36.6 kN/m
Design load for serviceability limit state	= 60.6 kN/m
Maximum bending moment at serviceability limit state	= 615 kNm
Design load for ultimate limit state = 1.4×36.6 + 1.6×24	= 89.8 kN/m
Maximum bending moment at ultimate limit state	= 910 kNm

For the composite reinforced concrete section chosen the calculations for the ultimate resistance moment and for the maximum stress at the service load were given on pages 22–23. The calculations for the design of the shear connectors are given in chapter 3.

Composite construction

Dead load of web of beam	= 0.9 kN/m

Characteristic dead load	= 34·7 kN/m
Design load for ultimate limit state	= 87 kN/m
Maximum bending moment at ultimate limit state	= 880 kNm
Area of steel beam	= 11 380 mm^2
Force in beam at full yield 11 380 × 0·93 × 0·25	= 2640 kN
Effective width of slab 9/5	= 1·8 m
Depth of compression zone at ultimate limit state	= 122 mm
Lever arm between compressive and tensile forces = 640 − 232 − 61	= 347 mm
Ultimate resistance moment 2640 × 0·347	= 920 kNm
Design strength of 19 × 100 mm studs = 0·75 × 100	= 75 kN
Number required to resist full yield force	= 35
Total number of studs in beam	= 70

Reinforced concrete

Characteristic dead load	= 37·1 kN/m
Design load for ultimate limit state	= 90·4 kN/m
Maximum bending moment at ultimate limit state	= 915 kNm
Area of longitudinal reinforcement	= 4910 mm^2
Design yield force 0·87 × 0·41 × 4910	= 1750 kN
Effective width of flange	= 2·1 m
Depth of compression zone	= 70 mm
Lever arm 640 − 73 − 35	= 532 mm
Ultimate resistance moment 1750 × 0·532	= 930 kNm

Serviceability

Reinforcement percentage	= 0·41
Permissible span/effective depth 20 × 1·38 × 0·8	= 22
Actual span/effective depth	= 16

DESIGNS B (see Fig. 1.15)
Composite reinforced concrete
Hollow precast units in lightweight concrete span between channels and there is an in situ topping of 75 mm. The slab is, from the dead load standpoint, equivalent to a solid slab of 230 mm thickness.

Dead load from slab $0.230 \times 23.6 \times 6$ = 32·8 kN/m
Dead load of web of beam   = 1·2 kN/m
Finishes and partitions   = 9 kN/m
Characteristic dead load   = 43 kN/m
Design load at serviceability limit state   = 67 kN/m

Maximum bending moment at serviceability limit state   = 680 kNm
Design load at ultimate limit state   = 99 kN/m
Maximum bending moment at ultimate limit state   = 1000 kNm
Area of steel channel   = 5883 mm^2
Design yield force in channel   = 1370 kN

Design ultimate force in prestressing strand   = 1120 kN
Total ultimate tensile force   = 2490 kN
Effective width of slab   = 2·1 m
Depth of compression zone at ultimate limit state   = 100 mm
Lever arm from centroid of channel   = 423 mm

Lever arm from centroid of strand   = 405 mm
Resistance moment provided by channel   = 580 kNm
Resistance moment provided by strand   = 455 kNm
Ultimate resistance moment of beam   = 1035 kNm

As the depth of the compression zone is not greater than 1/5 of the overall depth, there is no need to check the strain at the level of the strand. Nevertheless the calculation is given.

Strain at level of strand $0.0035 \times (405 - 50)/100$ = 0·012

This strain is just adequate to enable the full characteristic strength to be attained.

Serviceability

Approximate lever arm at service load   = 421 mm
Total steel force 680/0·421   = 1600 kN
Total steel area   = 6627 mm^2
Mean stress in steel   = 240 N/mm^2
Maximum stress in channel   = 250 N/mm^2
Studs required to develop yield force in channel   = 18
Total number of studs   = 52

Composite construction

Maximum bending moment at ultimate limit state	= 890 kNm
Force in steel at full yield 14 980 × 0·93 × 0·25	= 3480 kN
Depth in compression zone	= 145 mm
Lever arm of forces 500 − 157 − 73	= 270 mm
Ultimate resistance moment 3480 × 0·27	= 920 kNm
Number of studs to resist yield force	= 46
Total number of studs in beam	= 92

Reinforced concrete

Maximum bending moment at ultimate limit state	= 904 kNm
Area of longitudinal reinforcement	= 7365 mm^2
Design yield force	= 2630 kN
Depth of compression zone	= 104 mm
Lever arm 500 − 97 − 52	= 351 mm
Ultimate resistance moment 2630 × 0·351	= 920 kNm
Permissible span/effective depth 20 × 1·05 × 0·8	= 17
Actual span/effective depth 9/0·403	= 22

Thus this design in reinforced concrete, apart from leading to an unacceptable congestion of reinforcing bars, would not be acceptable from the deflexion standpoint. However, the greater effective depth of the steel in the composite reinforced concrete design does lead to an acceptable span/effective depth ratio.

APPENDIX 2

Calculations of hogging resistance moments of test beams

ASAAD'S BEAM 2A (equilibrium method)

This beam was reinforced on the tension side with seven 20 mm HT bars for which the measured 0·2% proof stress was 420 N/mm². The steel channel was 152 × 89 mm and of yield stress 280 N/mm².

Yield force in reinforcement 2199 × 0·42   = 922 kN

Assume depth of neutral axis to be 64 mm.

Area of channel in compression 152·4 × 7·1 +
    56·9 × 11·6 × 2   = 2400 mm²
Force in channel at ultimate load 2400 × 0·28   = 670 kN
Area of concrete in compression 152·4 × 64 − 2400 = 7360 mm²
Force in concrete 7·36 × 0·9 × 35   = 232 kN
Total compressive force   = 902 kN

The agreement between the tensile and compressive forces is sufficiently accurate for the purpose of calculating the lever arm.

Centroid of compression forces is 28 mm from top face.

Lever arm of forces 300 − 28   = 272 mm
Ultimate resistance moment 922 × 0·272   = 250 kNm

ASAAD'S BEAM 2A (compatibility method)

Measured strain at level of reinforcement   = 0·009
Stress in reinforcement at failure   = 450 N/mm²

Force in reinforcement 2199 X 0·45	= 987 kN
Measured depth of neutral axis	= 50 mm
Estimated depth of centroid of compression 50 X 28/64	= 22 mm
Lever arm 300 − 22	= 278 mm
Ultimate resistance moment 987 X 0·278	= 275 kN

NAJMI'S BEAM S2F

This beam was reinforced on the tension side with eight 20 mm HT bars for which the measured 0·2% proof stress was 480 N/mm². The beam also had two 20 mm HT bars acting with the channel in compression. The channel was 127 X 64 mm and of yield stress 270 N/mm².

Yield force in tensile reinforcement 2510 X 0·48	= 1200 kN
Yield force in channel 1898 X 0·27	= 510 kN
Yield force in compression reinforcement	= 300 kN
Area of concrete within the channel	= 5604 mm²
Force in the concrete within the channel 5·6 X 0·9 X 29	= 146 kN
Total compressive force	= 956 kN

There is a discrepancy between the calculated tensile and compressive forces, the indication being that the beam is 'over-reinforced', i.e. the neutral axis is somewhat deeper than the depth of the channel. Strains measured during the test indicated that this was in fact the case, but only 6 mm deeper than the channel. Since this small extra depth of concrete could not have withstood the extra compressive force required to balance the full tensile force, it is clear that other errors in the calculation must exist. These errors have been discussed in the discussion of the series 2 tests. The ultimate resistance moment is calculated below assuming a 'balanced' section. In design partial safety factors would be included.

Ultimate resistance moment based on tension 1200 X 0·28	= 336 kNm
Resistance moment due to compression reinforcement	= 82 kNm
Resistance moment due to channel force 510 X 0·286	= 146 kNm
Resistance moment due to contained concrete	= 40 kNm
Ultimate resistance moment based on compression	= 268 kNm